AF606799

Rediscovering Flavor

Alexander Wieck Fjaeldstad • Thomas Hummel
Robert Pellegrino

Rediscovering Flavor

A Practical Guide for Smell Loss

Alexander Wieck Fjaeldstad
Flavour Institute
Aarhus University
Aarhaus, Denmark

Thomas Hummel
Smell & Taste Clinic
Technische Universität Dresden
Dresden, Germany

Robert Pellegrino
Monell Chemical Sciences Center
Philadelphia, PA, USA

ISBN 978-3-032-08055-4 ISBN 978-3-032-08056-1 (eBook)
https://doi.org/10.1007/978-3-032-08056-1

This Springer imprint is published by the registered company Springer Nature Switzerland AG
The registered company address is: Gewerbestrasse 11, 6330 Cham, Switzerland

Acknowledgments

Recipe Authors

Christian Bøjlund
Aarhus, Denmark

Beetroot Hummus, Mango and Chili Salsa, Sun-Dried Tomato Pesto, Tzatziki, Classic Cheese Spread, Vegetable Sticks and Chips, Crispy Chickpeas with Spices, Stuffed Dates with Goat Cheese and Walnuts, Mini Pizzette, Fruit and Berry Skewers, Spaghetti a la Bolognese, Avocado Toast with Fried Egg, Buttered Cauliflower with Hazelnuts, Chimichurri, and Feta Spread, Meatballs in Curry with Spicy Rice, Indian Dal with Spiced Rice, Plaice with Curry Remoulade and Beetroot Coleslaw, Danish "Frikadeller" with Potato Salad, Quesadillas with Chicken and Tomato Salad, Beet Tartare, Portobello Carpaccio with Garlic Crostini, Jambalaya with Trinity Pickles, Smash Burger with Pungent Mustard, Lentil Casserole with Coconut Milk, Baked Sweet Potato with Filling, Quinoa Salad with Grilled Vegetables, Vegetarian Burger with Crispy Salad and Homemade Fries, Banana Split, Pancakes with Fruit and Syrup, Chocolate Mousse with Fresh Berries, Yogurt Parfait with Granola and Berries, Fruit Salad with Mint and Lime

Robert Pellegrino
Monell Chemical Senses Center, Philadelphia, PA, USA
Texas A&M University, College Station, TX, USA

Avocado and Lime Cream, Spinach Dip, Cold Spicy Nuts, Tangy Musubi, Olive-Spread Toast with Burrata, Çılbır with Breadcrumbs; Cocktails: Long Forgotten Friends, G & T, Cool Velvet, The Big Green, Bloody Buddy, A Very Lemony Sgroppino, Modern Gold Rush, Dominick the Donkey (Nonalcoholic), Hibiscus Mint Cooler (Nonalcoholic), Tamarind Dream (Nonalcoholic)

Franklin Finney
Public House, Knoxville, TN, USA
Dominick the Donkey (Nonalcoholic)

Recipe Authors

This cookbook was developed in close collaboration with individuals living with smell loss. Their lived experiences, sensory insights, and willingness to share their perspectives were invaluable in shaping the recipes in this book.

Danish Cooking Schools for Patients with Smell Loss

We are deeply grateful to all participants at the Danish cooking schools for patients with smell loss. Your feedback, enthusiasm, and culinary courage made these recipes better and more meaningful.

Personal Correspondence

We also thank the many individuals who shared their experiences, preferences, and suggestions through personal correspondence. Your contributions, though sometimes unseen on the page, are woven into every dish.

We thank "Aarhus Universitet Forskningsfond (Aarhus University Research Foundation)" for funding the illustrations.

Competing Interests The authors have no competing interests to declare that are relevant to the content of this manuscript.

General Introduction

The sense of smell plays a critical role in our perception of the world. Olfaction is crucially involved in behaviors essential for survival—identifying predators or hazards, recognizing individuals for procreation or social hierarchy, locating nourishing food, as well as creating an attachment between mating pairs, and between mothers and their babies. Similarly, odors are closely tied with memories and emotion, and our past experience of an odor can shape our perception of it. Finally, and most significant to the quality of our everyday lives, smell is the dominant contributor to flavor perception, having a profound effect on the human behavior of eating and sharing food. The inability to smell therefore removes many of the joys of cooking and eating.

Smell loss is not uncommon. Around 20% of the population exhibits an impairment of the olfactory system, with varying severity from total loss (anosmia) to partial loss (hyposmia) to deception in odor quality (parosmia, phantosmia). The quarter of the population who experience such loss are typically left feeling isolated, as friends and family fail to fully understand their hardship.

The field of chemical senses, which includes smell, but also two other senses (taste and irritation), is a rapidly changing and growing area of science. It is an interdisciplinary field, rooted in, but not limited to, chemistry, biology, medicine, and psychology that endeavors to explain how and why we experience odors and related multisensory sensations such as the flavor of food. In this book, we integrate the most recent science in olfactory research and flavor perception. We hope that understanding the scientific underpinnings of eating will offer a renewed possibility of experiencing the enjoyment of food to those without smell.

Flavor is not just smell, but a combination of many sensations that can be tweaked to enhance the eating experience. This knowledge has been essential for the development of recipes included in this book and, similarly, is essential for those wanting to create recipes of their own. These chapters have been written for those with smell impairment and for those with an interest in how each of our senses contribute information on food as we eat. This book is designed as a guide for individuals attempting to cope with their condition and provides them, and their immediate community of family and friends, with information about their disorder and the tools to make life enjoyable.

The *first part* begins with a description of how smell can fail us, explaining the common disability of smell loss. Next, we provide a primer on a healthy olfactory system (responsible for the sense of smell), explaining its anatomical structure as well as our own experience of smell, both its complexity and its bearing on daily life and the ways we interact with the world. *Part II* sets out to characterize the entire flavor system with a special emphasis on the olfactory network and will give the readers an understanding of the other chemosensory functions (taste and irritation) and their contribution, along with vision and hearing, to the multisensory experience of flavor. The objective is to provide an understanding of the underlying processes in order for the reader to have a framework for how flavor is built, allowing them to develop an approach to pleasure and enjoyment of food by recruiting all aspects of flavor apart from olfaction. Specific compensatory strategies to smell loss and eating are discussed, including sensorial and social aspects of eating. Described principles will be demonstrated with practical exercises, featuring tailored flavor kits. Next, focus is once again on smell loss and impairment, discussing treatments and protocols to introduce the concept of the unique plasticity of the olfactory system and olfactory training. We discuss important topics to patients, including how smell is measured and loss assessed, how best to talk with a treating physician, and what a smell and taste clinic can offer. Afterwards, we introduce a *third, integrative strategy* of cooking. We believe cooking incorporates both the sensory and social strategies while providing activity-driven smell training. By engaging the senses with new techniques and knowledge of culinary reactions, someone can learn the real joy of cooking even without a sense of smell and to create novel dishes to stimulate the flavor system. We provide examples of tested recipes in the recipe section, which are constructed through a dialogue between chefs, researchers, and patients with smell loss. This book is inspired by years of working with patients in the clinic and the kitchen alongside chefs. Recipes and techniques in this book are for patients to acquire confidence in the kitchen and increase flavor. They are sourced from direct experience with researchers and chefs having dialogues with patients about food. This book is designed as a culmination of that knowledge. It is a handbook for individuals attempting to cope with their condition and provide them, and their immediate community of family and friends, information about their disorder and the tools to make life enjoyable.

Contents

Meet the Authors

Thomas Hummel, M.D. is head of research in chemosensory systems at the Smell and Taste Clinic, Department of Otorhinolaryngology, Technische Universität Dresden, Germany. His work spans clinical and research efforts on olfactory and gustatory dysfunction, including neurodegenerative-related smell loss. He has played a key role in developing widely used olfactory function tests, such as Sniffin' Sticks and taste strips. He received most of his education at the Departments of Pharmacology at the University of Erlangen-Nürnberg and the University of Iowa, and at the Smell and Taste Center, University of Pennsylvania. He also led the influential "Position Paper on Olfactory Dysfunction" in 2017. Before entering Medical School, he was running a restaurant in Amberg, Germany, for one and a half years, serving the second best Schnitzel in town!

Robert (Bob) Pellegrino, PhD, DSc is an assistant professor of sensory science at Texas A&M and an affiliated scientist at the Monell Chemical Senses Center, specializing in flavor perception, smell perception and loss. With doctorates in Food Science and Medical Research, his work spans multisensory integration in eating, smell disorder diagnostics, and olfactory perception. He helped develop the widely used Scentinel diagnostic test and collaborates with the national Smell and Taste Center on cooking for those with smell loss. He also serves on the leadership team for the Global Consortium of Chemosensory Research (GCCR). He also consults for many beverage and food companies on product development and does food pop-ups around Knoxville, TN, and Philadelphia, PA.

Alexander Wieck Fjaeldstad, M.D., PhD is a rhinologist and olfaction researcher working at the Department of Otorhinolaryngology, Aarhus University Hospital, Denmark. He co-founded the Flavour Clinic, Denmark's first clinic for taste and smell disorders, where he focuses on diagnosis and treatment. He is a co-founder of the Flavour Institute at Aarhus University. He is also a member of the Center for Eudaimonia and Human Flourishing at the University of Oxford. Since 2019, he has been running a project on cooking schools for patients with smell loss, where methods for regaining pleasure from eating and cooking despite smell loss was the main focus. The tested methods and recipes from this project have been added to this book. During medical school, he worked in a cocktail restaurant, where cocktails were paired with food to boost the sensory experience.

Part I
An Olfactory Primer

General Introduction

The loss of sense of smell is common—and with it, most of your ability to taste food and drink. Olfactory dysfunction has many possible causes, with age and nasal inflammatory problems being the most frequent ones. First, let us identify the types of smell loss, their causes, and potential therapies and treatments. Then we'll look more closely at why we smell and how your sense of smell contributes to the way you taste food and how other senses can compensate for some of the lost flavor caused by olfactory dysfunction. These methods and tools are incorporated in the recipes in the final part of the book.

Chapter 1
Smell Loss and Primer on Olfactory System

Contents

What Is a Smell?

To understand the sense of smell and how the sense can be lost, it is worth spending a few moments to unpack what a smell actually *is*. The sense of smell is a chemical sense, meaning that for something to have a "smell," an "odor," or a "scent," it must release a particular type of small chemical molecule. Inside your nose, specialized nerve cells with specialized receptors detect those molecules.

You need to know about both the smell molecules—called odorants—and the smell receptors. We'll talk more about odorants and how they help produce the sense of "taste" in later chapters. For now, think of them as tiny particles that must physically interact with the smell receptors inside your nose in order for you to perceive any smell.

How do we smell? The obvious answer may be "through the nose," and that's partly true. When we inhale, part of the air flow is guided to the upper part of the nasal cavity, where the odorant molecules within the air reach and activate our smell receptors. When people talk about smell, they usually mean the smells sniffed from the air externally, such as "the smell of a rose" or revolting "stench of sewage." This type of smell is referred to as an *orthonasal smell*.

A. W. Fjaeldstad et al., *Rediscovering Flavor*,
https://doi.org/10.1007/978-3-032-08056-1_1

When we eat or drink, however, the upper palate allows odorants to travel from within the mouth up the throat into the back of the nose, reaching the same smell receptors that are activated through orthonasal smell. Often, when people refer to the flavor of a food or beverage, it is *retronasal smell*, or internally perceived odor, which dominates the discussion. In this way, the olfactory system has not one but two senses of smell. In fact, the "ortho" stems from the Latin prefix meaning *straight* while "retro" means *back* or *backward*, thus "orthonasal" describes smells that come in straight through the nose, while "retronasal" refers to smells that come in more from the back of the throat or behind the nose.

To understand this duality of smell, consider two different professions that concentrate on smell in a similar, yet distinct way: a perfumer and a chef. A perfumer makes decisions through external perception, mixing odorants together and wafting the chemical concoction in front of their nose. They wouldn't dare put this odorant in their mouth for a secondary judgment for several reasons: nobody is eating their perfume, and most of the odorants in the mixture are neither tasty nor safe for ingestion (nor would you want to ingest them!). On the other hand, "tasting" as we cook is part of the process of creating delicious food, and a chef would never send out a dish without critically evaluating the internal representation through eating, experiencing the retronasal smell of the ingredients and completed dish. (We'll take a deeper dive into how important retronasal smell is to the flavor system in Chap. 3.)

Depending on the route, the same odor source may be experienced differently. There's something completely unique about the sense of smell: the same sensory organ, located at the top of the nose, can interpret the exact same smell in a different way depending on which route the smell took to get there. Thus, the route taken by an odor provides a "duality" of smell. It functions both for sensing objects in the outside world and for objects within the mouth.

After odors have activated the smell receptors on the olfactory nerve cells, signals are transported via smell nerve fibers through a finely holed bone plate (the so-called cribriform plate) into the skull. Traumatic movement of the brain inside the skull can tear the fragile fibers and disrupt the smell signals from reaching the brain. When this happens, smell loss can occur immediately or within weeks of the injury and can be partial or complete.

Note: The olfactory system is a mirrored one, in which odorants travel either up the left or the right nostril and are processed relatively independently until reaching higher-order circuitry (deep in the brain). A depiction of one side is seen in Fig. 1.1.

Just above the nasal cavity, between the eyes, the activation of receptor signals merge in an advanced brain structure—the *olfactory bulb*. The activation signals from millions of olfactory receptors are well organized, as signals from identical smell receptor types merge in approximately 3000 small spherical structures (glomeruli) near the surface of the olfactory bulbs. This enhances and simplifies the patterns of activation, allowing for the creation of a unique odor image for a given odor.

From the bulb, fibers convey signals to the primary brain area responsible for smell in the brain, the primary olfactory cortex. This area of the brain is closely linked to areas responsible for pleasure, disgust, and memory, which allows for a rapid determination if there is danger in our surroundings, if we are close to

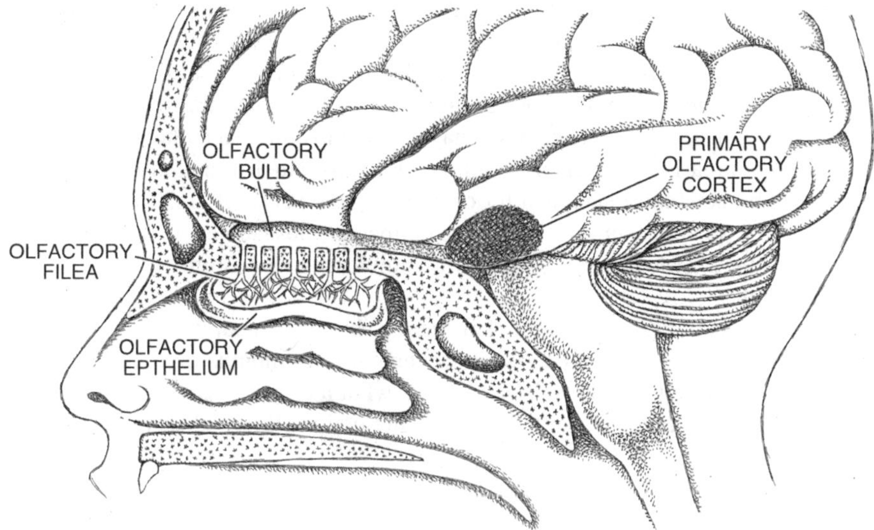

Fig. 1.1 Anatomical overview of nose, receptors, bulbs, and lower part of the brain

something edible, or if we should eat a given food in our mouth. Often, this system works without requiring much attention, especially if our focus is preoccupied elsewhere. However, if we eat something that does not meet the brain's predictions of the expected smell, we suddenly become more focused on the aromas of the food. This mechanism allows us to multitask while still protecting us from spoiled foods. As such, the processing of smell in the brain is closely linked to both pleasure and disgust.

Open Question: The Olfactory Bulb Is Not Needed to Smell?
The olfactory bulb is highly important for the perception of odors. However, it has been shown that there are people living without this key bulb who still maintain a normal sense of smell. In addition, we've seen evidence that people who probably never smelled in their life and have no olfactory bulb can learn to smell. Although we cannot disregard the importance of the olfactory bulb for the general population, this research shows that there are many open questions still unresolved in the sense of smell.

Hyposmia and Anosmia

People who are born with a functioning sense of smell can experience one or two of a few different types of symptoms related to olfaction: a decrease, an absence, or a form of distortion. Someone with an excellent sense of smell can slowly or suddenly become less sensitive to smells, or unable to detect any smells at all. *Hyposmia* means a decreased sense of smell, while *anosmia* means the complete absence of a sense of smell. It can be tough to tell where the dividing line is between a very faint sense of smell and a complete loss of smell. In addition, the threshold between a healthy sense of smell (or the technical name, "normosmia") and one reduced to the level of hyposmia is a matter of definition. Because smell function changes with age, varies by sex, and depends a lot on the way the sense of smell is tested, this definition of hyposmia can vary. Often it is defined as those lowest 10% of olfactory-test scores that are found in young, healthy individuals who claim to have a normal sense of smell.

It may be encouraging to know that there is hope for people with a decreased sense of smell. Many patients show improvement over months or years, and there are several effective treatments. We'll look at what a clinical evaluation looks like, and a variety of treatments for olfactory loss, in Section III (which covers recovery).

Parosmias and Phantosmias

Another category of smell-related symptoms involves distortions in the perception of smells. *Parosmia* means the distorted perception of an existing smell; just like with a healthy sense of smell, the experience of the parosmia will temporarily go away if you pinch your nostrils closed. Conversely, *phantosmias* are odor phantoms, which result from a person smelling something that is not actually there. Unlike with parosmia, you'll still be able to smell these odor phantoms even if you pinch your nose. Parosmias and phantosmias often occur together, and it can sometimes be difficult to separate the two.

Parosmias and phantosmias can be very disturbing to experience. It's often uncomfortable to perceive something that isn't there, or to eat something nice (and possibly expensive) and find that it tastes awful only to you. The distortions can be frequent. Thus, those suffering from them often get depressed by the constant reminder of their condition. This distortion is often not understood by others, and it's common for people with a healthy sense of smell to forget if their friend, colleague, or loved one has a parosmia.

The smell of a parosmia is typically unpleasant: it may smell like something is burning, rotten, chemical, or disgusting. Both parosmias and phantosmias are notoriously difficult to describe. Although "euosmias"—pleasant parosmias like the smell of "raspberry"—have been described, it can still be disturbing to perceive *everything* as smelling like raspberries. Cucumbers smell like raspberries! Fish smells like raspberries! Christine Kelly, who ran the UK-based charity *Abscent*, often describes a food shop as the Little Shop of Parosmia Horrors to parody the horror movie Little Shop of Horrors. Here, something that smells nice to most people, like a cup of coffee, smells disgusting to you (Fig. 1.2). Overall, parosmias and

Fig. 1.2 Parosmia alters the perception of odorants

phantosmias severely affect the quality of life. Below is just one of many accounts of parosmia from a patient:

> *"I seem to have two types of distortion 'categories'—coffee, chocolate, onions etc taste like a musky, nutty, rancid, earthy taste. And things like peppers and melon taste more chemical, like something that would be flammable".*
>
> –Parosmic Patient

Parosmias are typically found either in states of initial impairment or incipient recovery of the olfactory system. For patients who are about to recover from olfactory loss, a new parosmia is actually a sign of a slightly better or faster recovery. The majority of parosmias disappear within a year.

Odor phantoms are found in patients who recover as well as in those whose condition worsens. So, the diagnostic meaning of phantoms is less clear.

A small number of phantosmias also appear out of the blue and are associated with normosmia, typically in people in their thirties or forties. The presence of these phantosmias often varies with head movements and can be switched off by the so-called Valsalva maneuver (pinching the nose, closing the mouth, and then blowing up the cheeks and increasing pressure inside the mouth and nose). Unfortunately, this form of phantosmia may become worse over time.

Causes of Smell Loss

Congenital Anosmia

Around 1 in 8000 people is born without a sense of smell. It is rare for this condition to be detected within the first few years of a child's life. Often, the anosmic individual cannot recognize their own anosmia until they are somewhere between 5 and 15 years old, when the parents notice that their child does not react to foul smells as others do, or the child discovers that they cannot detect scents, which elicit reactions, be they positive or negative, from their peers. It can be difficult for doctors to determine with certainty whether the smell loss is congenital or acquired in early age, perhaps via one of the causes we'll discuss in the next few sections. It's important to pay attention when children experience smell loss before reaching puberty: congenital smell loss can be a part of Kallmann Syndrome, a condition characterized by delayed or absent puberty. Puberty can be safely medically induced, so an early diagnosis allows these children to enter puberty along with their peers. Currently, there is no treatment for congenital anosmia.

When people with congenital anosmia describe the flavor of food, they rarely use words related to smells/aromas but are often quite good at describing texture and basic tastes. When asked about the impact of their smell loss on quality of life, the negative effects are generally much lower in patients with congenital anosmia compared to those acquiring loss later in life.

Problems with the Nose and Sinuses

Because of the location of smell receptors, all the way in the upper part of the nasal cavity, airflow must have free passage to this part of the nose in order for us to smell anything. This may seem very simple, but there are many barriers and factors that play a role in determining which and how many odorants get to meet their respective smell receptors to create the sensation of smell. In fact, only 5–15% of the odorants that enter the nose actually reach this upper cavity, leading to a smell. This may sound like a lousy system, but this filtering of air is highly important to capture contaminants that might cause harm to a relatively open doorway to the brain (the sensory neurons). One broad reason people experience a reduction or loss of their sense of smell, either temporarily or in the longer term, is that this long, filtering nasal passage can become blocked, preventing smell molecules from reaching smell receptors. For example, many people have trouble smelling and tasting when they're sick with a common cold or are affected by allergies. This happens because the mucosa, or the soft tissue that lines the inside of the nasal cavity, becomes inflamed and swollen, blocking the pathway through the nose. We need for the air we inhale—and with it, tiny smell molecules—to be able to travel into our noses and all the way to the top of the narrow nasal cavity.

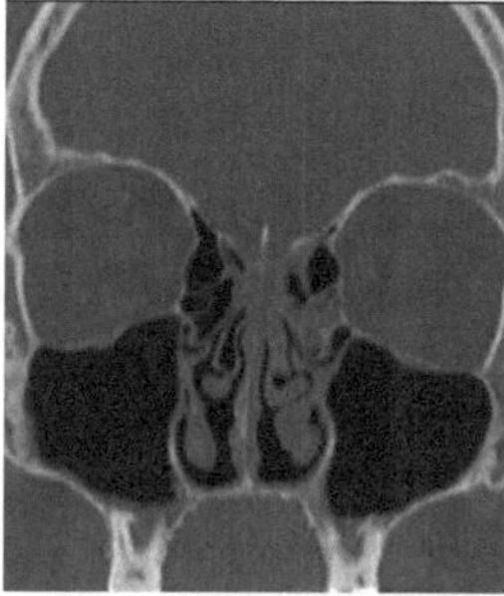
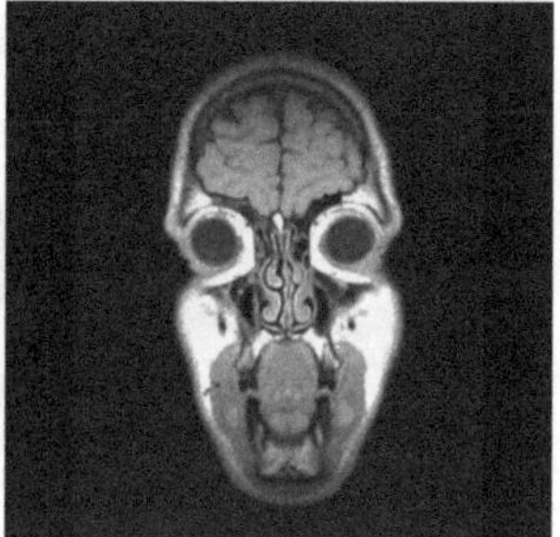
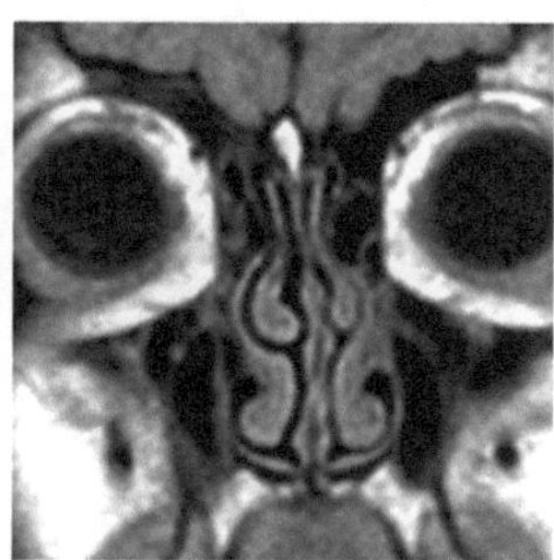

Fig. 1.3 An anatomical view with a CT scan and MRI of the nose. (**a**) A CT scan is a commonly used clinical assessment of the bones (white) and mucosa (grey) of the nose, where the airflow (air = black) is typically reduced in the upper part of the nose in patients with sinonasal smell loss. This patient has swelling in the upper part of the nasal cavity, causing a blockage of air and odorants to reach the olfactory mucosa. (**b**, **c**) MRI gives detailed information not only on bone but also on the soft tissue, as seen by the more detailed structure of the brain compared with the CT scan. This can be used to evaluate the olfactory areas of the brain

Typically, people experiencing a temporary illness or allergies recover from these symptoms of smell loss, nasal obstruction, and runny nose within a few weeks. However, for about 10% of the adult population, inflammation and swelling of the mucosa in the nose is chronic, with some even developing nasal polyps (e.g., swollen mucosa and increased mucus or pus) that worsen the blocking of the nose. Depending on the degree of swelling and anatomy of the nose, this can cause partial or complete smell loss. Smell loss due to physical problems with the nose and sinuses often fluctuates, which indicates that there is a well-functioning sense of smell hidden behind the swollen mucosae (Fig. 1.3).

Symptoms of nasal secretion or nasal blockage should be evaluated by an ear, nose, and throat specialist, while allergies can also be diagnosed and treated by allergy specialists. In some countries, ear, nose, and throat specialists are also responsible for diagnosing and treating allergies. With the proper treatment, the prognosis for olfactory recovery can be good (Fig. 1.4).

Smell Loss After Infections

Smell loss can occur after a wide range of different infections, but it's most frequent after viral upper respiratory infections such as a cold, flu, or COVID-19. These infections can cause swelling of the nasal mucosaS and/or secretions, which block the nasal passages and keep smell molecules from reaching the part of the nose that holds the smell receptors. Many people have experienced this "stuffy nose" feeling during a cold, finding themselves temporarily unable to smell as odorants cannot bypass the blockage. Infections can also cause damage to the smell receptors and underlying stem cells that are supposed to replace any damaged receptors. As a result, the infection can cause long-lasting damage to smell function, with sufferers

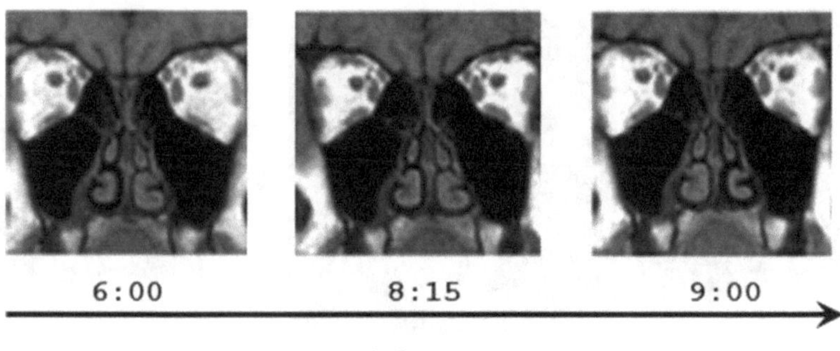

Fig. 1.4 The nasal cycle. The nose is more swollen on one side than the other. Every 2 to 4 hours, the sides shift, which can cause the sensation of unilateral nasal blockage that comes and goes. These MRIs on the same patient at different timepoints show how the swelling of one side of the nose shifts to the other side after a couple of hours

finding that smell does not recover even as other symptoms improve and disappear. Because the regeneration of these smell receptors is a slow process, patients often experience parosmia in the wake of these infections.

While persistent post-infectious smell loss has previously posed only a rare risk after other infections, COVID-19 has proven to take a particularly high toll on smell function after other symptoms have waned. Surprisingly, in contrast to other infections, young age does not seem to be protective against prolonged smell loss after COVID-19. This can be explained by the high affinity the virus has for the olfactory mucosa. Studies suggest that around half of COVID-19 patients suffer from both taste and smell loss, of whom half recover within the first few weeks. As such, many experience a prolonged smell (and taste) loss with a slow recovery rate after this specific infection. Although this has increased the number of patients with the heavy burden of living with smell loss, it has also drawn attention to smell disorders and prompted further research within this area. Hopefully, this renewed focus will result in an intensified effort for treating smell loss and alleviating the negative consequences of living with smell loss.

Head and Nose Trauma

Traumatic impact to the nose can result in a blockage of the nasal cavity, either due to a nose fracture or swelling of the mucosa. Normally, the symptoms will include physical alterations of the nose (e.g., deviated septum), nose bleed, swelling of the nose, or swelling inside the nasal cavity causing nasal obstruction. A displaced nose fracture should be realigned by a physician after 5–7 days, giving the trauma-induced swelling adequate time to go down. If not, the nose should be evaluated by an ear, nose, and throat specialist to assess if the displacement affects the nasal airflow and needs to be corrected with surgery at a later time.

Trauma of the head that does not include the nose can also cause smell loss, given the trauma is severe enough. Consider the case of a car crash, for example: just as any movable items in the car will be accelerated back and forth, the brain, too, will move back and forth inside the skull. This movement can cause damage to the surface of the brain, as areas involved in the registration and processing of smell can suffer damage and the small nerve fibers carrying olfactory signals from the nose can be severed.

Systemic Disease

A wide range of diseases and conditions can cause smell loss due to swelling of the nasal mucosa or affection of important brain areas processing smell information. These may be genetic disorders or conditions acquired through a person's life. Presently, there is no separate medical treatment for the smell loss itself, but only for the underlying condition. Adhering to treatment of these diseases decreases the risk of smell loss and may improve smell function.

In diabetes, commonly known manifestations include foot ulcers and affected vision due to nerve damage, but less focus has been given to the fact that about a quarter of diabetic patients exhibit smell loss. As this decreases the sensory input from food, it may affect dietary habits and blood glucose regulation.

For patients with reduced metabolism due to a low level of thyroid hormone, both taste and smell functions can be reduced. Reduced kidney function may result in decreased smell function and often also a distorted sense of taste. It seems, too, that many inflammatory disorders such as rheumatoid arthritis affect the sense of smell—not dramatically but to some degree.

Loss of smell may also occur in a wide range of neurological diseases, ranging from Parkinson's disease to multiple sclerosis. If you're experiencing a new taste or smell loss or distortion alongside other neurological manifestations (e.g., altered tactile sensation, muscle function, tremor, cognitive function), these symptoms should be evaluated by a neurologist.

Medications and Other Forms of Medical Treatment

A wide range of medications can cause smell loss (and even more frequently, taste disturbances—more on these in Chap. 3). This often happens in the weeks and months after the treatment is initiated but can also occur after longer periods of treatment. Some medications such as chemotherapy very frequently cause smell (and taste) loss, while others, such as antibiotics, pain medication, and medications against high blood pressure or cholesterol only very rarely affect the senses. However, as these medications are so widely used, even a very low risk can result in several cases of sensory loss (~ 2% of all cases). Patients who are able to stop using the medication that caused their smell loss, or substitute a different medication, often recover their sense of smell within weeks to months. For those unable to substitute medication, learning about smell loss and other working contributors to food (basic taste, texture, temperature, etc.) described later in this book will be vital to enjoying food. This is especially important as taste (e.g., sweet and salty) is often affected by medication in addition to smell.

Surgery of the nose entails a small risk of smell loss, while taste loss can occur after mouth, throat, and ear surgery. Some patients recover from these sensory deficits during the months after surgery, but the loss may be permanent. After nasal surgery, it is important to assess if swelling of the mucosa is causing the smell loss, as this can be treated medically.

Radiation therapy of the head and neck puts a strain on all cells but is especially hard on smell and taste cells. Furthermore, the production of nasal secretion and saliva can be affected, which can further inhibit smell and taste function due to drying and increased irritation (we'll explore the role that saliva plays in the flavor system in Chap. 4). Recovery after radiation therapy varies greatly. Improvement may occur over the course of years and often leaves the patients with some degree of combined sensory loss, brittle mucosa in the nose, mouth, and throat, reduced saliva production, and, in some cases, problems with swallowing.

Smoking, Toxins, and Alcohol

Although it is universally agreed that smoking definitely does not improve smell function, the extent of any negative impact on the sense of smell is much debated. Many patients experience an improved subjective smell function after they quit smoking, and smoking seems to increase the risk of smell dysfunction after head trauma. When smokers experience a new loss of smell, quitting smoking is usually recommended because cessation may improve their chances of recovering their sense of smell.

As a small proportion of all inhaled air reaches the smell nerve cells in the top of the nose, chemical toxins (e.g. solvents in paints) or metals (e.g. cadmium) can cause temporary or permanent damage to the sense of smell. The usage of facial masks and ventilation is essential in environments with exposure to these toxins.

Frequent alcohol use may cause damage to the smell nerves, as well as other nerves elsewhere in the body. This happens because alcohol can dissolve lipids (fats), including the lipid membranes that surround nerve fibers. This may affect both nerves in the limbs and more central nerves, such as the smell nerves. In chronic drinkers, brain regions involved in processing of smells may also be affected.

Age

In young, healthy adults, nerve cells with smell receptors cover the entire upper half of the mucosa in the nasal cavity. With age (often after 55 years of age), a partial shift occurs: an increasing number of cells responsible for processes related to

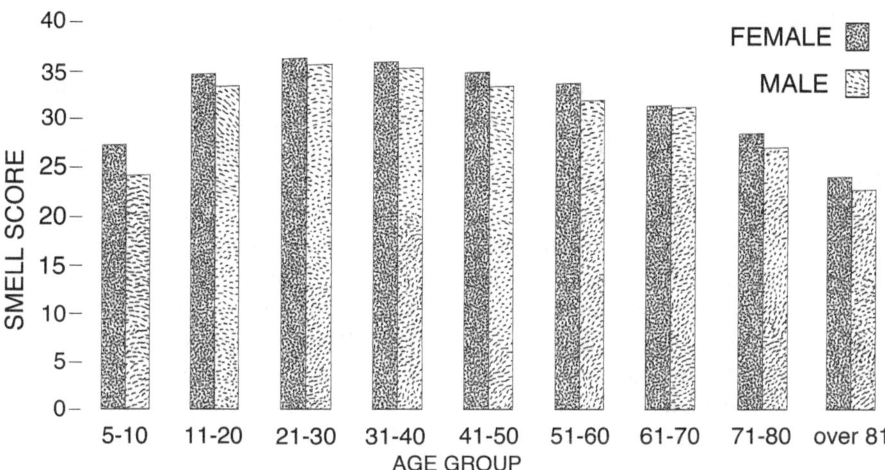

Fig. 1.5 Age-related changes in smell function

breathing begin to take the place of many of those smell receptors. As the surface area with smell receptors decreases, there is a decrease in olfactory function and an increase in risk for olfactory disorders.

As a result, large population studies on the sense of smell indicate that as people get older, their sense of smell begins to decline, along with similar impairments to the senses of vision and hearing. While the loss of these other senses can be reduced or fully compensated for by the use of glasses, surgery, hearing aids, or cochlear implants, no device currently exists for improving the sense of smell. While this can seem quite frustrating, there is some positive news about smell and aging. The French scientist Jean-Pierre Royet and colleagues have investigated how the sense of smell develops in a very particular subset of noses, namely in perfumers. Here, the sense of smell does not seem to decline with age. On the contrary, the volume of the brain areas related to smell actually increases over time (We'll look more closely at how the brain processes smell in Chaps. 3, 4, 5 and 6). With enough focus and training, the sense of smell can remain astute in spite of aging (Fig. 1.5).

Is There Hope in Olfactory Loss?

Many patients with olfactory loss show improvement over months or years! This is based on the regeneration of the olfactory neurons, and on the plasticity of the brain, especially the olfactory bulb. This improvement is especially pronounced in olfactory loss after infection, but also in patients with head trauma, and idiopathic causes (i.e., unknown causes). Smell improvement has even been shown in patients with Parkinson's disease. And most importantly, improvement of the sense of smell can also be found in older individuals.

The Olfactory System Can Regenerate After Injury

The olfactory system can recover from injury. This is in some part due to the regrowth of olfactory neurons in the nose. And this is unique. Such recovery is not found in the auditory system, for example. For those struggling with olfactory loss, the unique resilience of the sense of smell offers some hope. The regenerative capacity of the olfactory nerve should thus allow patients to reorient toward their future, with an understanding that recovery is possible. Depending on the cause of the smell loss and the particular individual, recovery can begin at different stages and proceed at different rates, but there are ways to increase the chances of a swift recovery. We will discuss these strategies in Part III, which covers recovery.

In general, the degree of improvement depends on various factors. Recovery is more likely if patients are of young age, have remaining olfactory function, exhibit parosmia, and are nonsmokers. Additionally, the smell loss becomes more resistant to recuperation the longer it persists. The person with the best chances for improvement is a young nonsmoking, healthy individual with olfactory loss for less than 1 year with relatively good remaining olfactory function and parosmia. Still, improvement can also be found, for example, in older people with low olfactory function without parosmia and olfactory loss for 2 years or longer. It is, however, less likely in these cases.

For cases of smell loss after viral infection, many studies suggest that recovery is frequent. Overall, between 30% and 60% of patients with long-standing olfactory loss improve over a period of approximately 1 year. Here it is important to note that improvement doesn't mean full recovery. Full recovery after a viral infection is less likely than improvement, but still, most patients with improvement are okay with their sense of smell.

Recovery rates are typically lower in head trauma than in post-viral olfactory loss. Improvement is found in 10–35% of the cases over periods of approximately 1 year. While on the one hand, it has to be emphasized that recovery is possible, the low recovery rate is somewhat disappointing. It is believed that the severity of the trauma is an important factor for the prognosis—the more severe the head trauma, the less likely recovery. Why is the rate of recovery in head trauma lower than in post-viral olfactory loss? It seems unlikely, given that damage and wound healing may produce scarring that can be an obstacle for regrowth of olfactory fibers. On top of that, the brain damage itself may not recover completely. That being said, numerous cases have shown that the sense of smell *can* return after a longer period of time. In one case, a 67-year-old woman exhibited improvement 7 years after the trauma. Though it seemed her sense of smell would never return, she saw notable improvement year 7 to year 9 after the accident. This case demonstrates the possibility of a late onset of recovery of olfactory dysfunction after head trauma.

Much of the recovery probably relates to the olfactory mucosa inside the nasal cavity. Here olfactory receptor neurons generate from so-called basal cells. The speed of recovery in humans is unknown, but in rats it is 2–4 weeks. Still, even when neurons recover at a high rate and high speed, it will take some time before they gain connections with the brain, and here specifically with the olfactory bulb. After all, some have to grow extensions (axons) over 6 cm—maybe more! Nerves, on average, grow 1 mm per day, and thus it may take 1–3 months before they reconnect again to the brain. The growth of these axons depends on age and numerous other factors.

Additional Reading

Clinical Olfactory Working Group consensus statement on the treatment of postinfectious olfactory dysfunction. Addison et al., J Allergy Clin Immunol. 2021 May;147(5):1704–1719. Online https://pubmed.ncbi.nlm.nih.gov/33453291/ or https://www.researchgate.net/publication/348472063_Clinical_Olfactory_Working_Group_Consensus_Statement_on_the_Treatment_of_Post_Infectious_Olfactory_Dysfunction. *Summarizes the careful opinion of an international group of clinical experts on established treatments of postviral olfactory loss.*

Olfactory Dysfunction in COVID-19: Diagnosis and Management. Whitcroft KL, Hummel T. JAMA. 2020 Jun 23;323(24):2512–2514. https://jamanetwork.com/journals/jama/fullarticle/2766523. *Summarizes current knowledge on diagnostics and treatment of olfactory loss.*

Cooper, Keiland W., David H. Brann, Michael C. Farruggia, Surabhi Bhutani, Robert Pellegrino, Tatsuya Tsukahara, Caleb Weinreb et al. "COVID-19 and the chemical senses: supporting players take center stage." *Neuron* 107, no. 2 (2020): 219–233. *Review of mechanisms behind covid-19 and the lost of smell, taste and chemesthesis*

Position paper on olfactory dysfunction 2023. Whitcroft et al., Rhinol Suppl. 2023 Oct 1;61(33):1–108. https://doi.org/10.4193/Rhin22.483. https://www.rhinologyjournal.com/Abstract.php?id=3097. *Detailed and extensive overview on various causes of olfactory dysfunction presented by an international group of medical doctors renowned for their work in the field of olfactory function.*

Olfactory dysfunction and its measurement in the clinic. Doty RL, World Journal of Otorhinolaryngology - Head and Neck Surgery 2015: 1(1):28–33. https://www.ncbi.nlm.nih.gov/pmc/articles/PMC5698508/pdf/main.pdf. *Detailed review on the assessment and management of various causes of olfactory loss presented by an internationally renowned, highly experienced clinical researcher in the field.*

Chapter 2
How Smell Works

Contents

Introduction

The human nose is quite a unique organ in the animal kingdom, a prominent facial bulge only surpassed by the long-nosed monkey. As we learned in the previous chapter, it is essential for breathing, where the mucosa of the small nasal cavity humidifies and heats the air before allowing it to enter the lungs. A part of the air flow is guided to the upper part of the nasal cavity, where it allows for small odorant molecules to activate smell receptors. This allows for us to register smells in our surroundings during inhalation and food aromas during expiration or swallowing. With an array of around 380 different types of smell receptors, we are able to distinguish millions of different smells. Upon activation of these smell receptors, the signal is within milliseconds transformed to an odor in our brain. This chapter outlines the processes behind this registration, how it is continuously replaced, and the mechanisms required to maintain the fine balance in the nose that ensures our ability to breathe and smell through this remarkable little organ.

A. W. Fjaeldstad et al., *Rediscovering Flavor*,
https://doi.org/10.1007/978-3-032-08056-1_2

How Smell Works

Chemoreception Versus Physical Sensation

It's important to note the different ways that external objects can interact with us. All information from the outside must interact with one of the five senses to be understood but can do so in a variety of ways. The senses of vision and hearing are often referred to as "remote" senses as they can obtain information from long distances via waves. The other senses can be thought of as "contact" senses as things must be nearby to provide relevant information. Information from these senses can come in physical or chemical form, each representing unique receptors. When someone grabs your hand, they are getting your attention by physically indenting your skin, while the temperature you feel from their cold hand is something quite different. The temperature sensation occurs through chemical contact, instead of physical, and is called chemoreception. Chemoreception can occur with all contact senses, which includes smell, taste, and touch. Interestingly, forms of contact can elicit the same response. Imagine getting a nose piercing, for example. You would be imagining a physically painful experience. But now think of eating a Carolina Reaper pepper (one of the spiciest peppers in the world). You'd also experience a lot of pain here as well; however, the pain would be from the chemical contact of capsaicin rather than the physical skin puncture. Odors are unique in that they can elicit two chemical feelings at once; consider mint, which is one part "minty" and another part "cooling." The "minty" comes from the sense of smell while the "cooling" comes from the trigeminal sensory system.

Why Do We Smell?

Why do we smell? For other senses, this may be an easy answer: we see to navigate, we hear to communicate, or we touch to hold onto objects. Appreciation for these physical senses can come much easier as they are tangible in a way that reflects our reality while more invisible senses that involve chemoreception may be more elusive. The physical senses represent an obvious utility for us today, being at the forefront of our attention; however, it was less useful to our water-dwelling ancestors in which the medium of water refracted light and broadened pressure, and for whom objects were not held. Here, the olfactory system, which creates the sense of smell, played a critical role in how the world was perceived which explains why it was the first sense to evolve. Smell enabled primitive species to make quick behavioral decisions based on low molecule chemical signatures in their surroundings. Noticing a particular odor tells us whether we should approach or avoid an object or situation, such as an encounter with a new creature or food. Smell today still helps us avoid or approach situations but also plays an important hidden role in many other aspects of life.

Most humans underestimate their reliance on smell in their everyday lives. Previous eras of scientists and thinkers, too, dismissed humans' sense of smell as inferior or insignificant. In contrast, most other mammals inherently understand the value of smell and rely on it for many of the demands needed to sustain life. Indeed, olfaction is crucially involved in behaviors essential for the survival of individual creatures, such as avoiding predators, recognizing other individuals for the purposes of procreation or social hierarchy, locating food, and bonding with mates. In fact, it turns out that humans also use their sense of smell for these same purposes (Fig. 2.1).

While doctors and scientists in general have a pretty good understanding of how the other senses work, you may wonder why knowledge of the sense of smell is generally lagging behind. While scientists began to describe the nerve cells in the

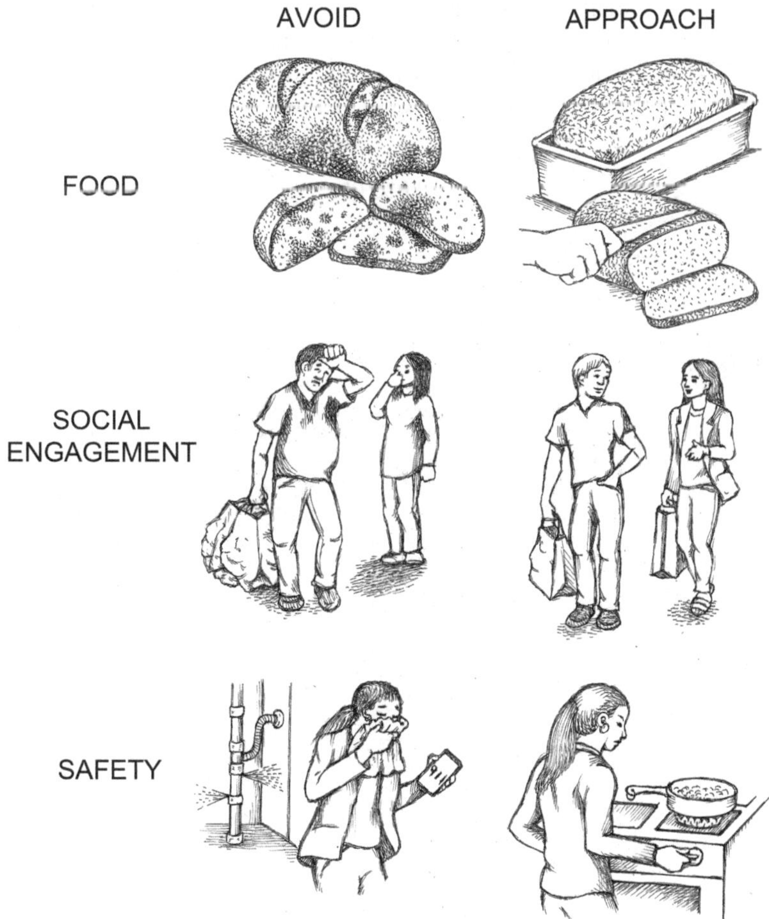

Fig. 2.1 Avoid and approach decisions guided by smell. Smell helps avoid spoiled foods and increases appetite and attention to foods with a pleasant aroma. In social engagement, smell plays a dual role: both in avoidance and in attraction. Smell can be important for alerting us in potential dangerous situations such as fires or gas leaks

eyes as early as 1894, the nature of olfactory perception remained an enigma until the Nobel Prize winning discovery of olfactory receptors by scientists Linda Buck and Richard Axel in 1991. This may very well be the reason why the sense of smell has often been referred to as "the forgotten sense," as both our understanding and the research conducted within this field has had a late start compared with our other senses.

What Is a Smell?

To understand the sense of smell and how it can be lost, it is important to first examine what a smell actually is. Smell is a chemical sense, meaning that for something to have a "smell," it must release small chemical molecules into the air. However, not all molecules are detectable by humans or animals, and only those that can activate smell receptors can truly be defined as smells (Fig. 2.2).

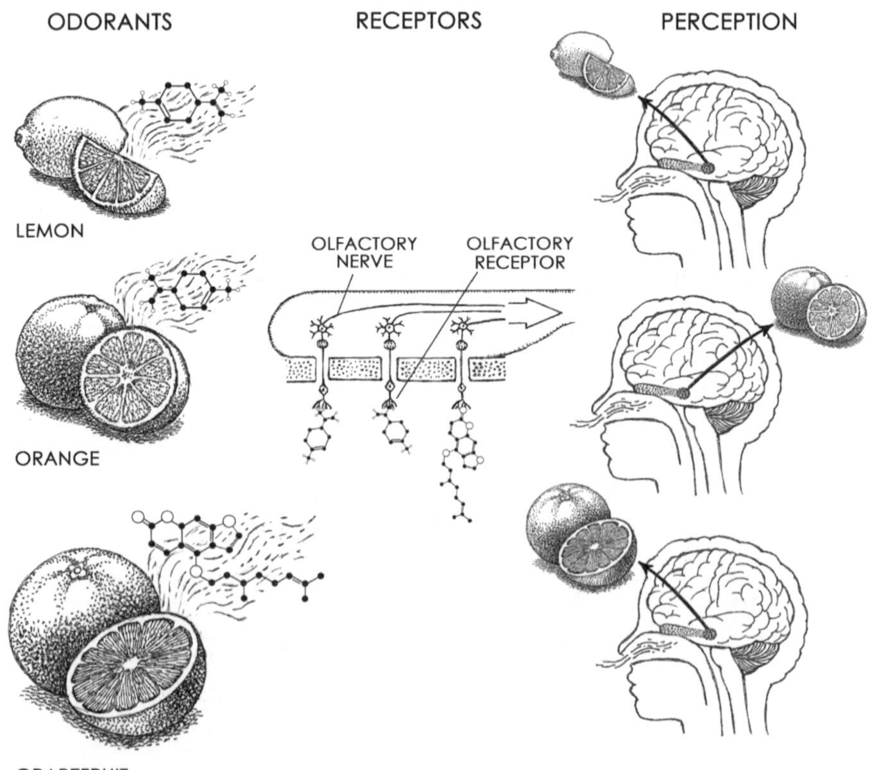

Fig. 2.2 The odorants (chemicals) bind to the olfactory receptors, allowing for perception of the food in the brain

Odorous molecules come in various shapes and sizes, but it is not their shapes alone that determine how they smell. For example, sucrose, a sugar molecule, is too heavy to travel through the air to reach olfactory receptors, so it does not have a smell, despite being detectable by taste. Interestingly, even trained perfumers or chemists may struggle to predict a smell based solely on the molecular structure, as molecules with similar shapes can smell different, and molecules with different structures can sometimes smell alike.

The discovery of olfactory receptors was a key milestone in understanding how we smell. The activation of these receptors by specific odor molecules can be compared to a barcode: each odor produces a unique pattern of receptor activation, much like how a barcode uniquely identifies a product. However, this "barcode" is personal, as the repertoire of smell receptors varies between individuals. For instance, some people have a specific receptor (olfactory receptor 6A2) that makes cilantro smell like soap, which explains why cilantro is a polarizing herb. This variation in receptors highlights that smell perception is both highly individualized and influenced by past experiences and familiarity.

To define what a smell is, it's crucial to understand both the odor molecules and the receptors they interact with. The relationship between molecules and receptors is often compared to a key fitting into a lock. However, unlike a key that fits one lock, an odor molecule can fit multiple receptors to varying degrees, activating some strongly while only weakly triggering others. Humans have around 380 types of active smell receptors, allowing for a vast array of receptor activation patterns. Most odors consist of a mixture of molecules in different concentrations, adding to the complexity of smell perception and enabling us to distinguish between millions of different smells (Fig. 2.3).

Journey of Smell Molecules Through the Nose

As smell receptors are situated in the upper part of the nasal cavity, airflow must have free passage to this part of the nose to facilitate the meeting of smell molecules and smell receptors. This may seem very simple, but quite often a blocked nasal passage inhibits the delivery of smell molecules to the part of the nose containing smell receptors.

The mucosa lining the nasal cavity has two primary functions: smell and respiration. To protect the lower airways, air is humidified and heated when breathing through the nose. This requires that the mucosa is able to adapt and adjust continuously for the ever-changing requirements, e.g., entering a heated room from the cold dry winter air or going in and out of air-conditioned buildings. Here, both temperature and humidity change abruptly. Heating of inhaled air is regulated by blood flow and swelling of the mucosa. If changes in nasal secretion do not immediately follow the demand, either the nose will start running or the mucosa of the nose and lower airways will become dry and irritated. Sensors are needed to provide detailed information to enable the nasal mucosa to adapt. There is actually an entire sensory

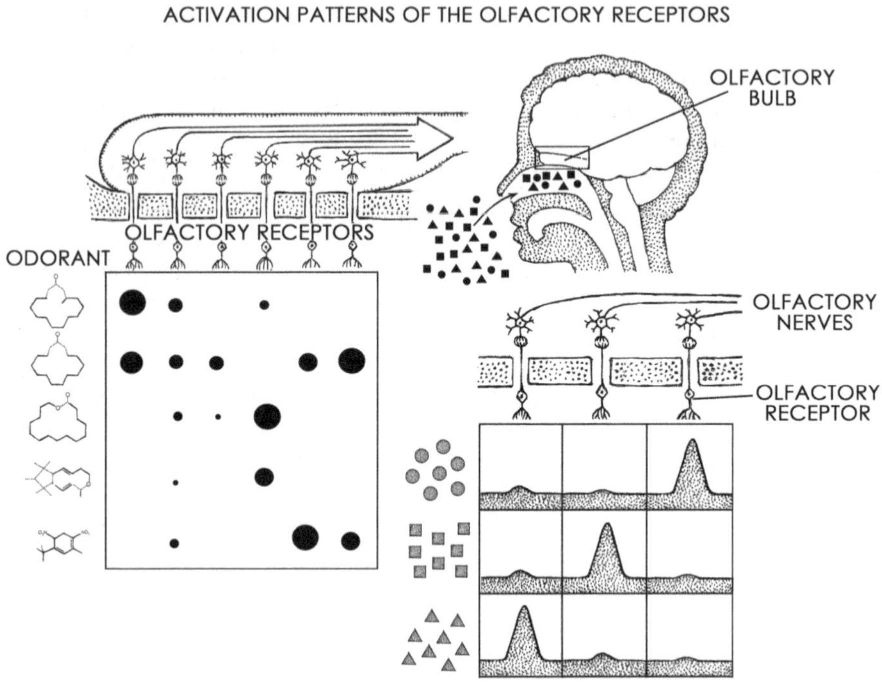

Fig. 2.3 Odorants typically activate several olfactory receptors with different strength, which creates a unique pattern of activation that can be recognized in the brain at later encounters

system located within the mucosa that provides information on temperature, pain, pressure, touch, and vibration. This is a part of the trigeminal sensory system (*nose feel)*, which also attributes important information to our overall perception of flavor, which is described in detail later in this book.

A part of this continuous regulation is the nasal cycle, where one side of the nasal cavity alternately will be more swollen than the other for 2 hours. This regulation of air flow and secretion can have large effects on olfaction, as the upper part of the nasal cavity is increasingly narrow toward the major area for smell mucosa in the very top, as seen in Fig. 1.4.

Smell in the Brain

After odors have activated the smell receptors on the olfactory nerve cells, signals are transported via small smell nerve fibers through a finely holed bone plate into the skull (*the cribriform plate*, as mentioned in Chap. 1). Here, just above the nasal cavity between the eyes, activation signals merge in two advanced brain structures, the left and right *olfactory bulbs*. The activation signals from millions of olfactory

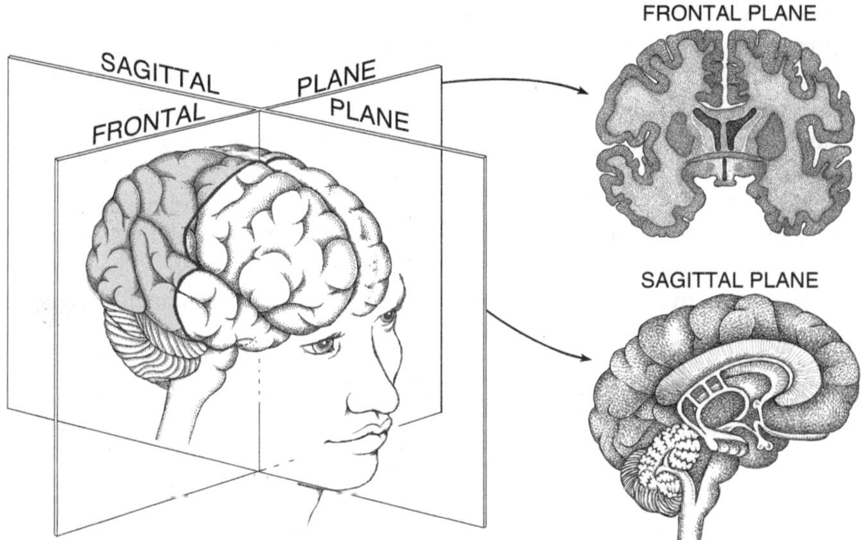

Fig. 2.4 Anatomical overview of the brain. Notice that illustrations can be made in different planes, as seen in the illustration of the sagittal plane and the frontal plane. These 2D planes are often used in books, scientific papers, and clinical settings to describe or examine brain structure or function

receptors are well organized. Signals from identical smell receptor types merge in small spherical structures called glomeruli, which help organize and simplify the activation patterns, creating the unique "barcodes" for each odor.

From the olfactory bulb, these signals are relayed to the primary olfactory cortex, the main brain area responsible for processing smell. This area is closely linked to regions responsible for emotions such as pleasure and disgust, as well as memory. This interconnectedness allows us to quickly assess our environment—detecting danger, identifying something edible, or deciding whether to eat a particular food. Often, this system operates subconsciously, allowing us to multitask while still protecting us from potential harm, like spoiled food. However, when there is a mismatch between the brain's prediction of an expected smell and the actual aroma, our attention shifts, making us more aware of the smell. Brain areas responsible for processing smell or other functions are often depicted in different planes as two-dimensional illustrations of the brain's three-dimensional architecture (Fig. 2.4).

Smell Memory and Imagination

Our sense of smell is deeply intertwined with memory and imagination. Unlike other senses, the olfactory system is directly connected to the brain's limbic system, which is heavily involved in emotion and memory. This close connection is why certain smells can instantly transport us back to a specific time or place, evoking vivid memories and strong emotions. For example, the scent of a particular perfume might remind you of a loved one, or the aroma of freshly baked bread might bring back memories of childhood. Smell memories are often more emotional and enduring than memories triggered by other senses, and these memories are created early in childhood.

The brain doesn't just respond to real smells; it also reacts to imagined odors. Research has shown that when we imagine a smell—like the scent of fresh-cut grass or cinnamon—the same brain regions that process real odors, such as the piriform cortex, become active. This suggests that our brains can simulate smells almost as vividly as if we were experiencing them. For example, simply reading the word "cinnamon" can trigger activation in the olfactory regions of the brain, even though no actual smell is present. The mere thought of a favorite food can make us hungry, while recalling a foul odor might make us feel queasy. Interestingly, if you ask someone to imagine a smell, they may even begin to sniff, as if the smell were real.

However, it's important to note that the ability to imagine smells is closely tied to our experiences. People who have never experienced certain smells, such as those with congenital anosmia (a condition where a person is born without the ability to smell), would find it difficult or impossible to imagine those odors. This suggests that our ability to imagine smells relies on our memory of past experiences with real odors.

Experiencing an Odor with a Food

Imagine standing just outside a bakery at dawn, moments before the door opens. As the door swings open, you're greeted by a rush of warm, fragrant air—a blend of yeast, flour, and caramelizing crusts. This experience begins with your nose, where odor molecules in the air enter through your nostrils. As these molecules travel through your nasal passages, some encounter the olfactory epithelium, a specialized tissue filled with receptors designed to detect these chemical signals.

When an odor molecule binds to a receptor, it triggers a neural response that travels through the skull base and into the brain's olfactory bulb. The olfactory bulb then organizes this information and sends it to the primary olfactory cortex, where the intensity and quality of the odor are evaluated. This region of the brain connects the scent with memories and emotions stored elsewhere, allowing you to instantly recognize the familiar smell of freshly baked bread, drawing you into the bakery.

Once you've bought the bread and hold it in your hands, you feel its warmth and notice the slight steam rising from its crust. When you take a bite, you hear the satisfying crunch of the crust breaking under your teeth. As you chew, aromatic molecules are released and travel up the back of your throat to your nasal passages, reinforcing the flavor. This internal perception of smell combines with the tastes on your tongue to create the full experience of flavor. Meanwhile, your brain coordinates sensory and motor processes—saliva production increases, breaking down the bread's starches and proteins, enhancing both taste and aroma. As you chew and swallow, the flavor and texture of the bread evolve, showcasing how our senses work together to create a rich and satisfying eating experience.

If you've never experienced odors (e.g., congenital anosmia), this scenario might be frustrating or even unimaginable. If you lost your ability to smell later in life (e.g., acquired anosmia), this scenario could evoke feelings of grief or nostalgia for something you can no longer experience. However, this example also highlights an important aspect of flavor—it's a multisensory experience. While smell plays a crucial role in flavor, it's not the only sense involved. Your experience with the bread is personal, varying among individuals based on past experiences, memories, and emotions. In the following sections, you'll learn more about the other senses that contribute to flavor and how, even without smell, these senses can be leveraged to enhance your enjoyment of food.

Additional Reading

Doty, R. L. (2015). *Handbook of olfaction and gustation*. John Wiley & Sons.

Buettner, A. (Ed.). (2017). *Springer handbook of odor* (pp. 459–482). Cham, Switzerland: Springer International Publishing.

Cobb, M. (2020). *Smell: A very short introduction* (Vol. 637). Oxford University Press, USA.

Olofsson, J. (2025). *The Forgotten Sense: The Fascinating Science of Smell with a Mind-Expanding Perspective*. Harper Collins Publishers.

Part II
Smell, Flavor, and Eating

General Introduction

Have you ever heard someone say *"This tastes like strawberries" or "This smells sweet"?* These are common ways of describing how we perceive food and are good examples of the large gap between how we *describe* food and how we *sense* food. A strawberry tastes sweet, but the aromatic flavor that makes a strawberry irresistible is in fact pure smell. Even so, we never perceive sweetness by smell, but by taste.

The senses are almost always intertwined and can be difficult to distinguish, both in the way we talk about food and how we perceive it. To explore and reap the full potential of our senses and the flavor of our food, a better understanding of both how the senses work and what our senses tell our brain is key. This part describes all sensory input that contributes to the flavor of food, which allows us to enhance the experience of eating—both with and without a normal sense of smell.

Chapter 3
Smell's Prominent Role in Flavor Creation

Contents

Smell's Prominent Role in Flavor Creation (Retronasal Smell)

Evolution of Retronasal Perception

Smell, along with the other chemical senses, evolved early in life's history, allowing organisms to communicate simple behaviors like *approach* or *avoid*. This mechanism helps organisms survive and thrive. Over time, species evolved, and new biological changes led to new sensory perceptions. About 250 million years ago, a significant anatomical shift occurred in some reptiles, later inherited by mammals and primates: the development of a hard palate (the non-moving roof of the mouth) and a softer one, creating a gap between it and the throat. Ultimately, this allowed smells to be perceived both externally and internally during eating, enabling the experience of flavor.

This dual sense of smell, using both orthonasal (external) and retronasal (internal) pathways, is essential for experiencing the full flavor of food. For instance, the enjoyment of strawberry ice cream or the complexity of a glass of red wine relies heavily on retronasal perception. Humans inherited this ability, enhancing our appreciation of food and contributing to our evolutionary success, particularly with

A. W. Fjaeldstad et al., *Rediscovering Flavor*,
https://doi.org/10.1007/978-3-032-08056-1_3

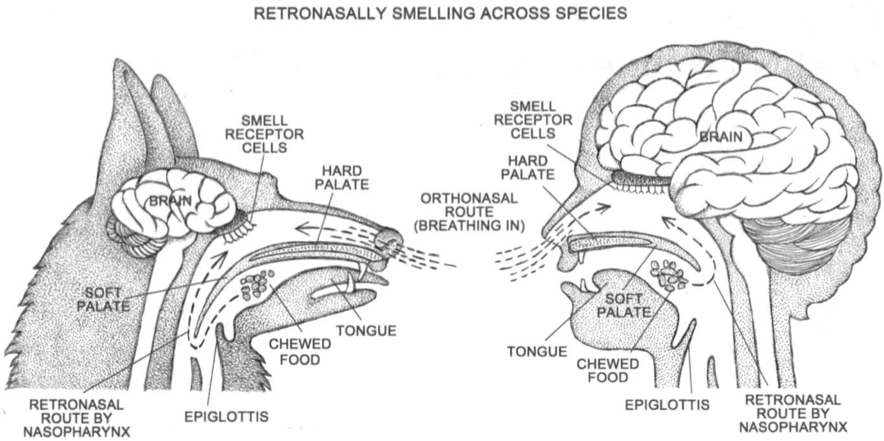

Fig. 3.1 The hard palate was formed in some animals 250 million years ago, which enabled nasal respiration. The length of the soft palate influences the volume of air that is pulsated into the nose during swallowing, resulting in retronasal smell and flavor creation. Human olfaction has a shorter soft palate and a less pronounced bony structure (not shown, but called transverse lamina), thus having less restriction of airflow from back of the throat to the olfactory smell receptors

the advent of cooking, which allowed us to process high-energy foods for our demanding brains.

Figure 3.1 illustrates that, unlike primates (such as humans), other mammals like mice and dogs—often celebrated for their superior sense of smell—have anatomical features that limit retronasal olfaction. Their soft palate extends to the junction of the esophagus and airways, restricting direct airflow between the mouth and nose and thereby reducing flavor perception during eating. Additionally, the transverse lamina, less pronounced in primates, is a bony structure that encloses the upper nasal folds and isolates the olfactory region. This structure further hinders air from the oral cavity from reaching the olfactory receptors, diminishing retronasal olfactory sensitivity in non-primates. In contrast, humans, with our evolved nasal anatomy and significant brain regions dedicated to sensory processing, excel in food evaluation. Retronasal smell is critical for perceiving flavor, while orthonasal smell enhances the eating experience by creating anticipation.

Duality of Smell

The traditional view, dating back to Aristotle, holds that humans have five senses: smell, taste, touch, sight, and hearing. However, more recent studies suggest we have many more senses, both external and internal, such as balance (equilibrioception), body movement (kinaesthesis), and temperature (thermoreception). These senses help us perceive and respond to our environment, often without our conscious awareness.

As we've established, we have two distinct types of smell: orthonasal and retronasal. For example, you can smell coffee by sniffing its aroma or taste it by sipping and breathing out through your nose, experiencing the flavor internally. This duality of smell, using the same sensory organ, allows us to perceive the same odor differently depending on its route into the nose. Experiencing perceptually different smells for the same odor source is not uncommon. The aroma that draws you to a coffee or cheese purchase may not perfectly match the experience once you take a sip or bite and perceive it from inside your mouth. Take Limburger cheese, for example—its scent from a distance is often compared to sweaty socks, yet once it moves past the nose and onto the palate, it reveals earthy, buttery, and even floral notes. While both perceptions remain undeniably cheese-like, the qualitative experience shifts due to differences in the odorant's path—whether through orthonasal or retronasal olfaction. Some striking differences between the two routes of smell are the retronasal route has an incorporation of saliva, a longer route with different air flows, and other sensory perceptions that often accompany internal odorous sensations.

For something to smell, it must reach the olfactory sensory neurons in the mucous layer of the upper nose, located between the eyes. Both internally and externally perceived odorants must hit this sensory region for smell perception. The palate allows odorants from the mouth to travel up the throat into the back of the nose, reaching the olfactory mucosa similarly to external sniffing. Once the sensory neurons are activated, the same cascade of activations occurs from the olfactory bulb to the brain. Sniffing helps external odorants reach these areas, while gulping pushes internal odorants up into the back of the nose.

Swallowing

You can test this gulping mechanism in action by holding your index finger in front of your nose and then swallowing. You should feel a gentle flow coming out of your nose and that is air from the that would carry any stray odorants in its path.

Food can be thought of as a molecular matrix with interweaving proteins, carbohydrates (or sugars) and fats that create its structure as well as its nutritional value. Odors may live on the surface of these food matrixes or be entrapped within them. When food enters the mouth and the jaws begin to fracture its structure, entrapped odors, as well as those on its surface, are freed and become volatile within the oral cavity. The volatility of an odorant is determined by its boiling point (or temperature at which they turn into vapor)—odorants with a low boiling point are more volatile. Odors rapidly go volatile within the mouth and nose as both are basically heat chambers typically above temperatures experienced outside the body. Remember, the inside of the mouth runs hot as you probably have noticed when using a thermometer (around 37 °C/ 98.5 °F). Thus, it's always summer in the mouth. Along with physical destruction from the teeth, the food matrix also encounters chemical deterioration as saliva is added to the mix.

Saliva!

Saliva is a watery liquid secreted into the mouth by several glands, providing lubrication for chewing and swallowing, and aiding digestion. It is composed mostly of water (98%) along with enzymes that help break down food. These salivary enzymes can also interact with odorants in the food, potentially altering their volatility and even their molecular structure, leading to the formation of new odor molecules with different smells. Consequently, the odorants reaching the sensory organ via the back of the throat can be quite different from those directly sniffed through the nose.

For example, consider M&Ms eaten straight from the fridge: they are cold, crunchy, and sweet with a very subtle orthonasal smell. As you chew, the heat and moisture in your mouth, combined with saliva, release new odors, transforming the texture from hard and cold to creamy and warm. These changes are influenced by individual variations in saliva, which differ among people and fluctuate throughout the day. The amount of saliva can increase due to homeostasis such as hunger or even just by watching an appetizing food video, highlighting how dynamic the role of saliva is in flavor perception.

Flavors Are Projected into the Mouth

When you experience an odor during eating, it often feels like these odors are being perceived in the mouth—although by now you now know that all odors, regardless of route, are processed in the nose. This phenomenon is referred to as odor referral and demonstrates the dependency of odors with other senses in the mouth. When eating, several senses are experienced together rather than temporally spaced out, making it difficult to separate the senses involved and leading us to discuss the experience as multisensory (e.g. flavor). There might be some indirect benefits to perceiving food odors in the mouth rather than the nose. For instance, to dispose of something rotten that you have in your mouth you'd want to spit it out rather than sneeze or snort it out. Multisensory perception (like the flavor of food) creates sensory dependencies in which manipulation of one sense impacts another. Retronasal odors are typically perceived with other sensations (e.g. taste, texture). Together, they represent oral percepts. The absence of one of these sensations may lead to cognitive interference (mis-match signals). Very low (even unconscious) or high levels of taste can impact retronasal experience smell and thus it's not hard to conjecture that internal smell representation depends on this sense.

Use Taste to Increase Smell

You can improve the ability to detect a smell by as much as 28% when combining it with congruent taste stimuli (sweet taste + cherry odor), even if this sweet taste is below a detectable concentration. The same does not happen with incongruent pairings (savory taste + cherry odor). This may be important for those with partial olfactory loss as it demonstrates a way to increase sensation.

The same applies for texture: moving your jaw and tongue alone can increase the intensity of smells. Looking at the language of smell reveals many of these dependencies. Compared to orthonasal perception, retronasal smells are discussed in more abstract terms that are less representative of the referred objects that are defined by vision (e.g., shape, color, size). For instance, we typically discuss things by their visual identity in western cultures and extend this description to smells externally perceived. For instance, a strawberry-like odor when presented in isolation may be referred to as "fruity" when smelt orthonasally, but more abstractly as "fresh" when consumed. This change does not seem to be related to the sometimes lower intensity (or strength) of internally perceived odors and thus may be due to less information being interpreted as other senses typically accompany it during the eating process (e.g. taste and texture of the strawberry). Neuroimaging studies back this idea as smells activate many more brain areas such as areas related to textures or tastes.

The Patients' Perspective

While the tongue has receptors for the five tastes (sweet, salty, sour, bitter, and umami), the nose contains over 380 types of receptors. On each of the 6–30 million olfactory neurons in the nose, the surface is coated with one single receptor type. Each receptor can bind to multiple, different odorants. This large possible combination of odor receptors and its binding properties leads to our ability to identify hundreds of thousands of potential odorants (this number is heavily debated). As we've discussed, the smells perceived by an odorant are route-dependent and have different functions with food. Milk beyond the expiration date is first smelt and then ingested when no smells indicating the milk has gone sour or rotten are detected. Sniffing a food also plays a modulatory role, helping to set taste and flavor expectation. The retronasal smell contributes a significant proportion of the experience of flavors for a majority of foods. Although there is a debate on the exact contribution of smell to flavor (most estimating 75%), there is an overall agreement that smell is a dominant player in our overall perception of food. This odor component provides several of the complexities of flavor that characterize a food such as the sulfur from blue cheese or the caramel of toasted bread. Similarly, the meaty, floral, fruity, herbal, citrus, and burnt notes all derive primarily from the contribution of olfaction.

To demonstrate this contribution to someone who can smell and to bring smell loss impact on flavor into perspective, the following exercise can help (see Fig. 3.2). This is a simple jelly bean test. Here, someone pinches their nose while chewing a jelly bean and then, after some 15 seconds of chewing, releases their fingers during the chewing process. Before releasing their pinched fingers, the individual will be unable to identify the flavor of the jelly bean, only experiencing taste qualities like sweet and sour.

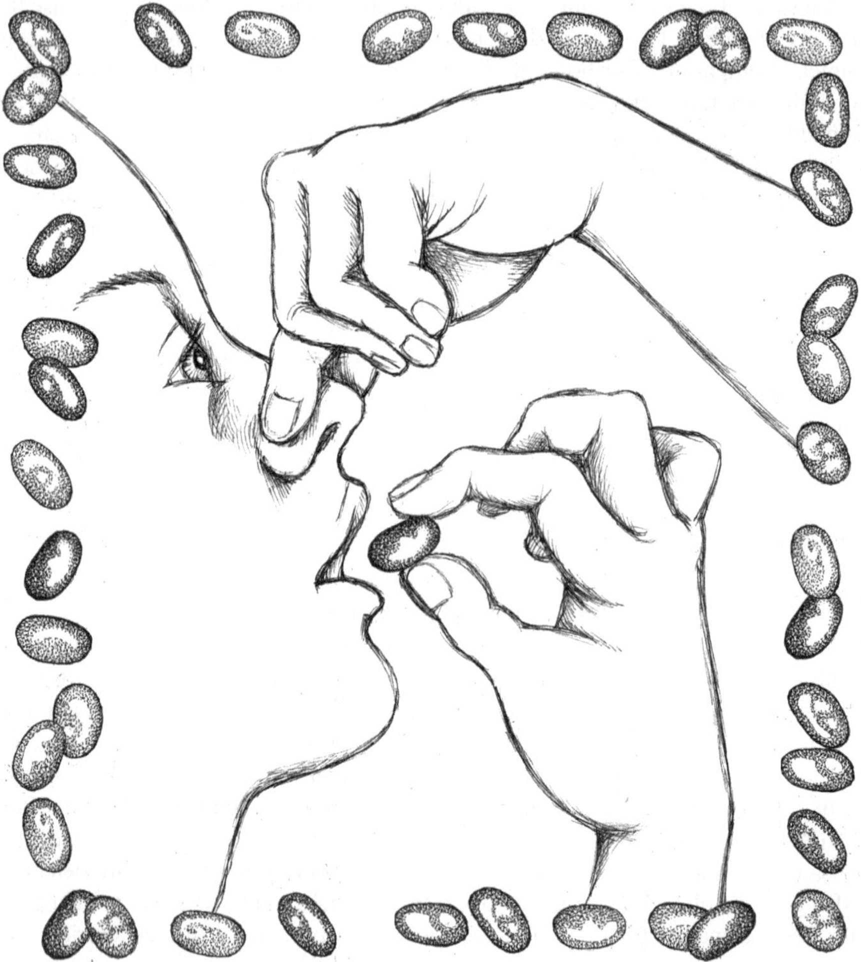

Fig. 3.2 The jelly bean test. Pinch your nose and chew a jelly bean (or similar candy). Sense the taste (sweet), and when you let go of your nose, sense the olfactory activation when the aromas reach the nose

However, upon release they will be immediately informed of the flavor. This exercise also helps to cast doubt on the inherent feeling that retronasal odorants are smelled in the mouth. Interestingly, the simultaneous presentation of odor through both routes (ortho- and retronasally) can even increase the strength of the odorant when eating

Manipulating Flavor with Orthonasal Smell

In light of the dominant role that smell plays in food perception and enjoyment, it is no surprise that several modern cooking techniques have manipulated and enhanced olfactory cues in many foods or dishware to increase liking and satisfaction of meals. For instance, atomizers are commonly used to induce another layer of flavor in foods and cocktails, while companies such as Molecule-R Flavors have introduced an "Aroma Spoon" that is designed to introduce new aromas to food with each bite. Even the simple trick of warming up food can enhance its flavor as odorants become more volatile with higher temperatures as mentioned earlier with the volatility of molecules in the mouth (a natural heat chamber). Someone might have used this trick to disguise day-old bread with a quick bake in the oven to make it smell fresh again. If water is added to the bread, the crust even becomes crisp, and the mouthfeel and sound will add to a perception of freshness. This knowledge, and application of techniques, can be used to enhance the experience of flavor for those with recovered or partial loss of smell; however, total loss of smell must rely on incorporating cues of other senses rather than solely enhancing smell (Fig. 3.3).

Fig. 3.3 Orthonasal olfactory boost with an "Aroma Spoon"

The loss of orthonasal smell often accompanies the loss of retronasal smell as they activate the same sensory organ. This impacts the many enjoyments of eating as the dominant player in flavor is gone. Without intervention or insight, eating may become more of a refueling of energy rather than a pleasurable experience. Here, patients sometimes report noticing their hands shaking as a reminder they had not eaten for some time. These patients also report increased social anxieties while dining out. If you don't have a sense of smell, you know this situation all too well. If not, imagine sitting at a table with friends/family when the food comes out. As everybody begins to eat, they start to critique and analyze the flavors they are experiencing with the food, "What is that herb?"; "That's delicious." They might even ask your opinion about the food, "Do you like it?" or "What does it taste like?". As someone who cannot smell, you would have trouble relating, as the eating experience would be much different from others at the table. You might even feel inadequate or embarrassed for not being able to have the same experience or respond appropriately. These feelings remain in the kitchen, unsure of how the food tastes when cooking and could even be a safety issue when spoiled foods (that often give off a cautious aroma) may not warrant a look at the expiration date. For individuals without a sense of smell, many warning signs and flavor complexities are lost. Therefore, flavor takes on a new form where other senses must be given more weight or a new perspective on the eating experience must be taken to feel the joys of eating or cooking.

Food Fact: Heating and Chopping Food

Just like the aromas of the food are released by heating and mastication in the mouth, these methods can be used to amplify aromas in the kitchen. Smells are chemical compounds, which are released in much higher numbers when the temperature increases. For patients suffering from a partial smell loss, several techniques can be applied to compensate for the reduced sense of smell. Roasting or baking vegetables, bread, or meat releases a lot of flavors. During heating, carbohydrates and proteins of the food interact in a chemical reaction called Maillard reaction. Apart from giving the food a brown color, crispy crust, and notes of sweet caramel, the reaction releases an array of intense flavor compounds. Cutting—or perhaps even muddling—fresh herbs into a dish/salad can also increase the intensity of the smell. Often, these intense aromas are liked by persons with normal smell too, which makes it ideal for cooking a pleasurable meal for the whole family.

Summary

Smells are perceived both through the nostrils (orthonasal smells from the external environment) and through the back of the nose (retronasal smells from the mouth and internal environment). The same smell can elicit very different sensations and experiences depending on which route it enters the nose. This knowledge can be used to enhance the flavor of a meal and can be exploited to improve the eating experience.

Additional Reading

Books

Bojanowski, V., & Hummel, T. (2012). Retronasal perception of odors. *Physiology & Behavior, 107*(4), 484–487.

Ni, R., Michalski, M. H., Brown, E., Doan, N., Zinter, J., Ouellette, N. T., & Shepherd, G. M. (2015). Optimal directional volatile transport in retronasal olfaction. *Proceedings of the National Academy of Sciences, 112*(47), 14700–14704.

Shepherd, G. (2013). *Neurogastronomy: how the brain creates flavor and why it matters*. Columbia University Press.

Articles

Pellegrino, R., Hörberg, T., Olofsson, J., & Luckett, C. R. (2021). Duality of smell: Route-dependent effects on olfactory perception and language. *Chemical senses*, *46*, bjab025.

Rowe, T. B., & Shepherd, G. M. (2016). Role of ortho-retronasal olfaction in mammalian cortical evolution. *Journal of Comparative Neurology, 524*(3), 471–495.

Rozin, P. (1982). "Taste-smell confusions" and the duality of the olfactory sense. *Perception & Psychophysics, 31*(4), 397–401.

Small, D. M. (2012). Flavor is in the brain. *Physiology & Behavior, 107*, 540–552.

Spence, C. (2020). Multisensory Flavour Perception: Blending Mixing Fusion and Pairing within and between the Senses. *Foods, 9*(4), 407. https://doi.org/10.3390/foods9040407.

Chapter 4
Introduction to the Flavor System

Contents

Introduction

We use the word *taste* in a number of ways: from the ability to discriminate quality, to having an urge to savor a particular food, to our actual physiological *sense* of taste, responsible for the perception of sweet, sour, salty, bitter, and umami. The most common—but, in fact, misleading—way is our use of the word *taste* to describe the fundamental perception of food's flavor in our mouth when we eat.

There are actually numerous senses involved in forming that perception of food—a complex interaction of the senses—involving smell, taste, vision, touch, and sound. A good meal can become even better if plated in a visually appealing way. A lukewarm soda gives us only a fraction of the pleasure an ice-cold soda will have on a summer day. A potato chip without the accompanying sound of *crunch* while chewing will disappoint. And eating during a common cold, when our stuffy nose deprives us of aromatic input, that absence of smell—the dominant contributor to our perception of flavor—makes for an empty, vacuous experience. Our perception of food is truly a multisensory experience. To begin to orient ourselves toward how the senses interact, we'll describe this experience in terms of *flavor*.

A. W. Fjaeldstad et al., *Rediscovering Flavor*,
https://doi.org/10.1007/978-3-032-08056-1_4

The human brain is a fascinating thing. Aside from consciousness, emotions, memories, and dreams, this complicated organ has the ability to register a wide array of information from our surroundings. While a smartphone can sense images and perform facial recognition, the eyes constantly record visual input from our surroundings and conduct immediate analyses of not just faces but also colors, patterns, and all other known objects in close collaboration with memories. The sense of touch is not restricted to pure contact but gives you information on surface structure, temperature, pain, and vibration. The hearing organ in the inner ear registers sound waves in an astonishing spectrum of different frequencies that are translated into meaningful words, music, or danger signals. The sense of smell constantly analyzes the presence of hundreds of thousands of different small molecules in the air, while taste receptors in the mouth provide detailed information on the chemical properties of foodstuff in the mouth.

As a carefully conducted symphony, all these senses give their input to the brain during food consumption to avoid spoiled foods, ensure adequate nutritional intake, and increase the pleasure of eating. In this multisensory waterfall of peripheral sensor activation and brain processing, the different sensory organs are important instruments when decomposing a meal into how you actually perceive the ingested food.

Even though the brain is perhaps the greatest supercomputer to ever exist, we would be overwhelmed if we had to consciously register all of these signals.

This is where attention and predictions come into play. Imagine driving your car to work. Most people use the same route and have automated all the small constituents of this relatively complex—yet easy—procedure. Try and remember what happened during any given regular trip just a few days back— often not an easy task, as the procedure is automated and without much conscious thought or formation of memories. This is because the brain saves energy by predicting what will happen. If something unexpected happens, a prediction in the brain is not correct, and we have to behave accordingly, bringing the sensory information to our consciousness—if we see a new billboard, if there is a scratch in the car, or if roadwork forces us to differ from the regular routines and the related predictions. This mechanism is identical for food consumption, where we are perhaps even more alert—any small changes in the sensory signals from the food may indicate that the food is spoiled. From an evolutionary perspective, sickness from spoiled foods would have impaired our ability to hunt or protect ourselves. However, in the case of disorders of the sensory apparatus, the change in sensory signals is not related to altered food, but the predictions of sensory input in the brain are changed nonetheless.

The Multisensory Contributions to Flavor: All Senses Contribute

Of all the senses, smell is generally considered the strongest contributor to the flavor of food. Nevertheless, the important role of the other senses is emphasized when eating chips that have lost crunchiness, eating a slice of apple that has turned brown,

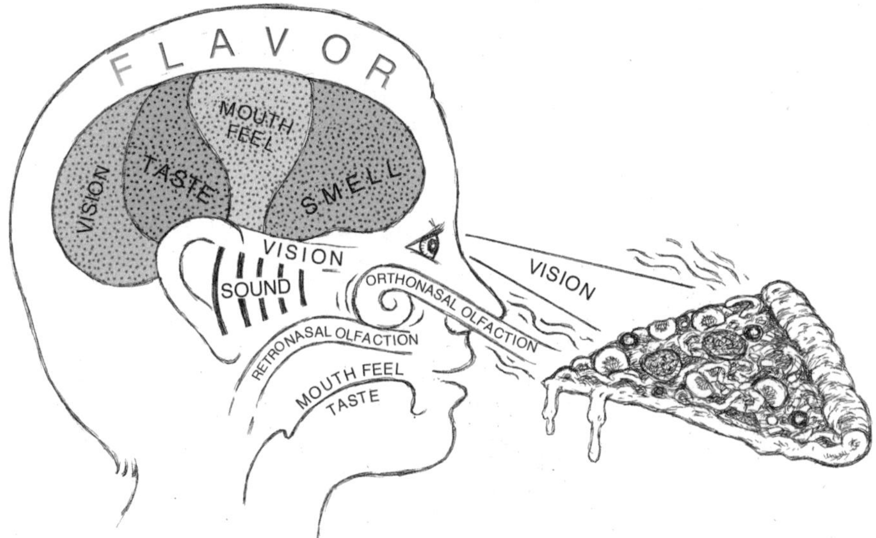

Fig. 4.1 Multiple senses are activated during eating

having an egg without salt, or being served a warm coke or sauce béarnaise with lumps. In these situations, the other senses take center stage in flavor perception (Fig. 4.1).

When our expectations of the flavor, temperature, or texture are not met, these become the predominant sensations. What if we could use this knowledge as a stepping stone to improving the experience of food without olfactory dominance? By exploring the other senses, we can create new expectations and experiences for creating and enjoying food. This chapter will give an overview of how the other senses work, how the mechanisms of compensating for other forms of sensory losses differ from compensating for a smell loss, and how many sensory instruments we actually have in our flavor toolbox that are simply waiting to be applied in recipes and take center stage (Fig. 4.2).

Taste

The term "taste" is typically generalized to describe the overall flavor of a food, such as "this tastes good," but in fact, it is only one part of the equation. To use this generalized notion infers that the food not only tastes good but also smells good, looks good, feels good, and sounds good when chewed in the mouth. True taste comes from sensory nerves within tiny taste receptor organs called taste buds scattered on the surface of the tongue, palate, and throat. There can be up to 4600 taste buds on the human tongue. Taste buds are arranged in macroscopic structures called

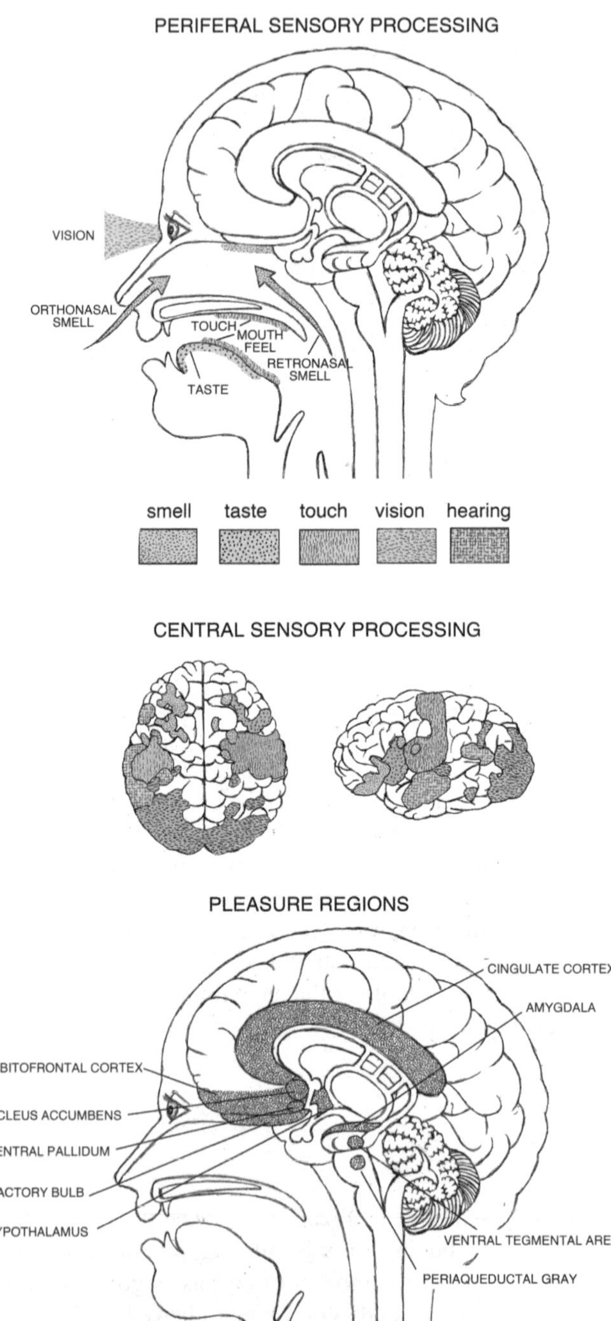

Fig. 4.2 Anatomy of the senses and pleasure regions in the brain. The brain regions of the brain related to memory and evoking feelings of pleasure and disgust are located in close proximity to the region of olfactory processing. These areas are highly connected and responsible for the effects smell can have on our behavior and memories

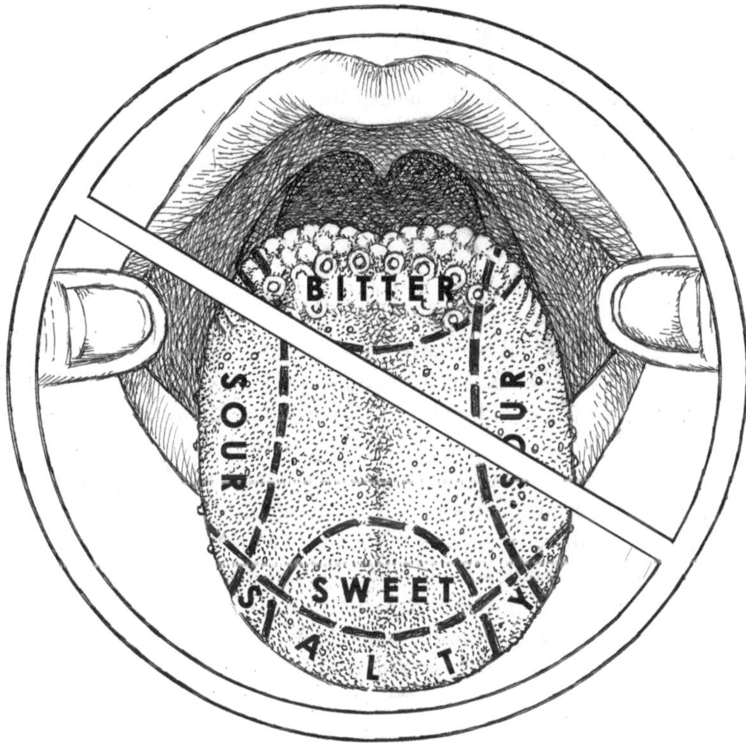

Fig. 4.3 The tongue map

taste papillae, typically containing 50–100 taste receptor cells. If you hold your tongue out in front of a mirror, you'll see a ridge of bumps that are a type of taste papillae. Although taste buds in different locations on the tongue are visually distinct, they all contain the same taste receptor cell types, enabling the perception of all basic tastes across the whole oral cavity (Fig. 4.3).

"The Taste Map"

An incorrect popular belief of taste has been the "taste map," which states that specific areas of the tongue or palate are most responsive to one particular taste. This idea of a taste map originated from a paper written in 1901 titled *Zur Psychophysik des Geschmackssinnes* by David Hänig. In this extremely detailed work, he showed differences in perceptual thresholds of tastes across the tongue and palate. Results over the years underwent a bad game of telephone until these sensitivity differences were misconstrued as a difference in sensation. However, while differences in sensitivity may exist across the

(continued)

tongue, all parts of the tongue can experience all tastes (except for the middle of the tongue!). To prove the textbooks wrong, all somebody had to do was touch a sugar cube to the side or back of the tongue and notice its sweetness was present there just like the tip of the tongue. In fact, taste cells are everywhere (also in the gut—'gastrointestinal chemosensation")—although we cannot perceive the taste of food past our throat, our gut can detect nutrients and respond accordingly.

Yet the life of the taste bud is relatively short-lived, usually spanning 2 months. After an injury, new taste buds are formed after 10–14 days. As food enters your mouth, enzymes quickly begin breaking it down into a dissolvable form that saturates the homes of the taste buds, leaving you with a sensation of one or many tastant(s). The five basic, or fundamental tastes are sweet, salty, sour, bitter, and umami. The latter has been acknowledged as a basic taste for more than a century in Japan but lacked formal recognition in the Western world until around the 1980s. To this day, many have difficulties identifying and describing the umami taste in the Western world—not least exploiting the richness this taste can add to a meal during cooking. Each has biological importance that helps us determine what to ingest and not ingest, such as nutrients or toxins (Fig. 4.4).

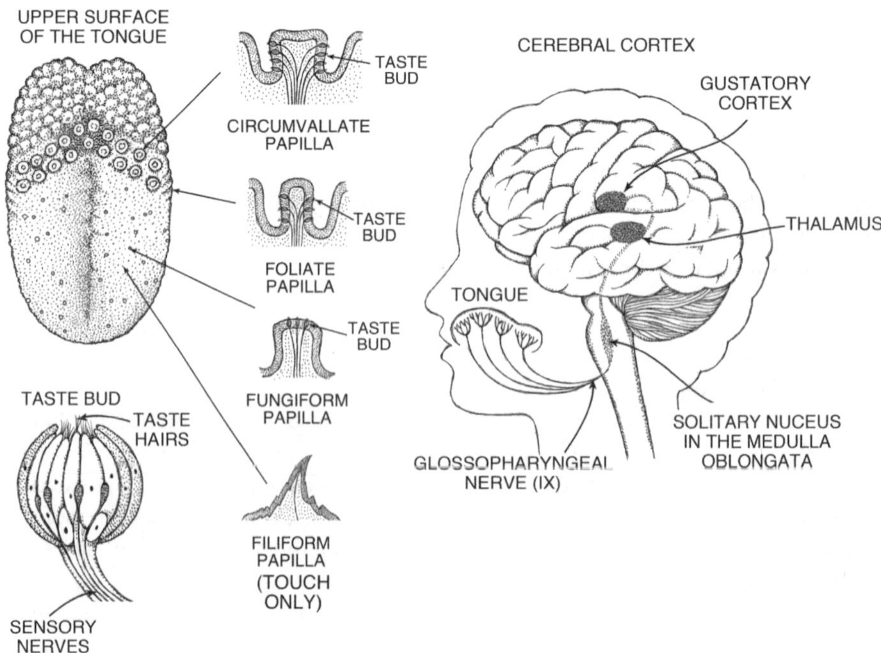

Fig. 4.4 Taste bud, taste papillae, and tongue diagram

The body has evolved mechanisms to detect particular components in food to either accept or reject it and, if the former, balance the intake of the food. Five basic tastes have developed, each associated—in principle—with the detection of a particular food component: sweet with energy, salt with electrolytes, bitter with toxins, sour with acids and ions, and umami ("savory") with amino acids/proteins. From birth in both humans and rodents, the patterns of reaction are strikingly similar: sweet stimuli elicit a licking, seeking behavior, while bitter stimuli elicit aversion. The evolutionary nature of these five basic tastes can be seen across species with the simple notion "use it or lose it." For instance, bitterness is a common taste in plants; thus, you would expect herbivores, who only eat plants, to have less sensitivity to bitter and enjoy plants more. This is exactly what nature has done with herbivores (plant eaters) perceiving less bitterness than omnivores (plant and meat eaters) followed by carnivores (meat eaters). Similarly, the domestic cat is strictly a carnivore, and thus does not need to have a sweet receptor. Functioning as gatekeepers, the taste buds are ready to analyze the chemical components of the oral cavity, whereby swallowing of inedible and potentially toxic items can be avoided. In addition to this innate defense, the sense of taste functions in a complicated interplay between reflexes and learning, between perception and expectation, and between pleasure and disgust.

The perception of taste and smell is closely linked, as described in the chapter on retronasal olfaction. From a neuroanatomical perspective, the two senses are directly linked through neural fibers. This enables modulation of sensory activity at an early stage in brain processing, resulting in intertwined signals already before the area of conscious sensory processing is activated. In other words, when we have difficulties differentiating smell from taste, the wiring of our brains is the perfect excuse.

Vision and Flavor

While four-legged animals still use their noses to identify potential sources of food, the transition to walking on two feet has distanced our noses from scent trails and shifted the sensory importance to rely more on information from the eyes to investigate if a given food is edible. We depend on vision to conduct a preliminary evaluation in order to decide if the food should be picked up, brought closer to the face for evaluating the smell, and perhaps be placed in the mouth, where texture and taste are evaluated before swallowing. As such, vision is key in activating attention that can end up allowing a given food to enter the fragile gastrointestinal system with all of its attendant dangers, i.e., digestion of spoiled foods, toxins, or bacteria. The changing of colors of fruits and the shine of the skin are important cues when we determine if fruits are ripe. This is often exploited in supermarkets who prepare foods in order to make them more visually appealing—apples are made more shiny, and meat is packed to preserve the red color.

In order to simplify all of the things we see, we depend on our brains to conduct some rather advanced pattern recognition. We quickly learn to recognize faces and to interpret facial expressions, which is important to function in a social context. In the same way, we continuously learn to link different kinds of foods with positive or negative eating experiences. The mere sight of a Danish or a steak may awaken the appetite for some and instantly increase the production of mouth water. Others may have stronger reactions when looking at oysters, lobsters, strawberries, or something completely different. In this function as the initiator of choosing the food and eating, vision seems to have great potential for both changes in attention and for setting up expectations for the subsequent perception of flavor.

These expectations depend on prior experiences and can be widely different between cultures and even individuals. If a Briton is presented with a red soda beverage, he/she will anticipate a cherry or strawberry flavor, while a blue soda will cause anticipation of a raspberry flavor. In contrast, Taiwanese expect the red soda to have a cranberry flavor and the blue soda to have a minty flavor. These expectations can alter the overall pleasure experienced during consumption; a red-colored beverage with strawberry flavor is generally rated as tastier when compared with a green-colored beverage with the exact same flavor. The color preferences may change as our knowledge and experience grow; while the color brown is often a sign of decayed raw vegetables and meat, the brown color of cooked food is perceived quite differently. When food containing both amino acids and sugars is heated to around 140–165 °C (280–330 °F), a chemical reaction occurs (the Maillard reaction). In this process of caramelization, the surface of meat, bread, or vegetables turns brown but also results in a crispy crust, a sweet caramelized taste, and numerous favored aromas, as mentioned in Chap. 9. As a consequence, the brown color is suddenly perceived as a positive contribution to the expected tasting experience, as long as we expect the color to originate from the preparation.

Our prior experiences have tremendous influence on our perception of food, where a mismatch in expectations and any sensory input can give rise to a significant reduction of pleasure during eating. As this mechanism applies to all senses, an analogy between vision and smell may help people with a normal sense of smell to understand the devastating effects smell disturbances may have on enjoyment of food. Imagine eating an apple where you can spot signs of decay. This can be compared with the sensory input for some individuals with parosmia, which is a distorted perception of odors often found in patients with olfactory loss.

Experiment

Blood orange and yellow beetroot. When presented with a jelly made of orange and beetroot, people naturally assume that the orange is orange and the beetroot is red. However, chef Heston Blumenthal has been known to trick the senses of his restaurant guests. In this simple dish, he served jelly made of red blood orange and orange beetroot and gave the following instructions: “An orange jelly and a beetroot jelly, I suggest you start with the orange.” It took a while for the guests to figure out that their expectations of colors and flavors had been inverted.

Vision is not limited to altering the expectations of flavor; vision can actually affect how we perceive input from our other senses. If you, for example, add a color to an odorless liquid, the illusion of an odor is produced for people smelling it. Color can modulate the experience of an existing smell, as the addition of color alters smell intensity. The visual impact on flavor perception is not restricted to the color of the food but also the color of the container in which it is served. Research studies have shown this effect in several different experiments: coffee tastes more intense and less sweet when it is drunk from a white mug rather than from a clear glass; desserts taste up to 10% sweeter when served from a white container rather than a black container; soup served in a blue soup plate is perceived as being saltier. It is time to explore how visual manipulations of the surroundings, table, tableware, presentation of the dish, and garnish can shift flavor attention from smell to vision.

Hearing and Flavor

For many, sounds are not something they often relate to eating, but hearing can have large effects on flavor perception in multiple ways. A lot of sounds are linked with food, directly or indirectly. This has been exemplified by the work of psychologist Ivan Pavlov more than 120 years ago. He found that by coupling a food stimulus with the sound of a bell, he could quickly get a dog's mouth water to run merely by ringing the bell. We are often made aware of the presence of food by more related sounds, which may suddenly remind us that we are either hungry or thirsty (e.g., sizzling sounds during preparation of food or sounds related to the packing, opening, or pouring of the food or beverage).

Hearing is also directly involved in eating, e.g., chewing, masticating, slurping, biting, or sucking. Through the bones of the head—and in particular bones close to the inner ear, like the jaw—vibrations can easily reach the inner ear. Therefore, the act of chewing causes vibrations, which leads to the perception of a sound even if it cannot be heard by other people. Imagine recognizing the sound of a bag of potato chips being taken from the kitchen shelf—in many situations this will have an immediate effect on attention, appetite, and saliva production for these crunchy, salty potato chips. When placing the salty flakes in the mouth and chewing, the crispy crunch gives immediate pleasure, and it can be hard to leave the bowl until it is completely empty. However, if the chips have gone soft and the crunchy sounds are missing, the cravings can disappear just as fast as they emerged. The same mismatch in expectations can be found in many other types of foods and beverages, e.g., when opening a bottle of sparkling wine or sinking your teeth into a crisp apple. As such, the expected sounds related to flavor can play an important role in the overall perception of flavor and pleasure (Fig. 4.5).

Background noise can also deeply alter the perception and pleasure of a meal. Imagine you are at an Italian restaurant where subtle Italian music blends in with the background noise. It seems authentic and can enhance the overall pleasure of the native dishes. On the other hand, loud, unrelated background noise can suppress our

Fig. 4.5 Texture as a multisensory concept. The crunch of a crisp biscuit or apple can have an essential role in the perceived freshness and pleasure

ability to perceive the flavors of food. Noise may even cause distortions of flavor. It has been shown that noise can selectively alter the perception of taste, as the ability to detect sour and sweet can become impaired while leaving perception of others unaffected. Some types of noise can also reduce sensitivity to odors.

Recent research has even shown that sounds can be directly related to other senses. Bitter taste is thought of as having a lower pitch, while sour and sweet have a higher pitch when people have to match sounds and taste. As such, the congruence of sounds is more complex than using geography to match the meal and song: By playing classical music instead of pop music in a cafeteria, people were willing to pay more for the same food; by turning up the volume of background music, people were buying and drinking more beverages and eating their meal faster (even people eating/drinking alone); by playing a certain type of music in the supermarket, the tendency to buy certain wines changed; by playing pleasant sounds as compared with neutral sounds, subsequent odors were perceived as more pleasant. All of the abovementioned effects happened subconsciously, which emphasizes the link between hearing and flavor. Accordingly, as sound can affect other sensory input, attention to music and noise during a dinner may aid in regaining pleasure from a meal if other senses are affected.

Mouth-Feel: Touch, Texture, and Temperature

How We Feel What Is in the Mouth

Have you ever spent several minutes attempting to remove a piece of food between your teeth with your tongue, only to be surprised by the ridiculously small size of the responsible vegetable or meat? We are extremely sensitive when it comes to our oral cavity due to a combination of high receptor density and an array of different receptor types enabling us to obtain highly detailed information about anything in the mouth we could consider swallowing.

While touch sensors give information on the size, shape, and texture of the food, touch alone cannot explain all sensations of food. Pressure sensors add another dimension to the mere sensation of touch. Pain and temperature sensors contribute with a wide array of information, from a mechanical bite in the tongue to chemical stimulation from chili or peppermint and the recognition of hot and cold. Taken together, the sensory nerve is responsible for registering touch in the entire face and also contributes with highly sensitive information on how the food feels in the hands, mouth, and nose.

Touch input comes in many forms, supplementing different types of information, and these change throughout the eating experience. By containing different types of sensors, differentiated and accurate information on touch, pressure, pain, and temperature in the mouth is registered. This adds to the evaluation of the food, where viscosity, texture, temperature, pain, and irritation are evaluated in the brain and compared with expectations. Although little thought is often given to the contribution from this sense, the broad vocabulary often used to describe the properties of food may ignite a more apt appreciation. Think of all the different words you use to describe mouthfeel, e.g., astringent, burning, cooling, creamy, crisp, dry, fatty, hard, hot, juicy, metallic, pungent, sticky, soft, starchy, tart, tingling, or watery. In fact, in restaurants the most common sensory source of complaints is the mouthfeel: i.e., the chips are soft, the wine is too warm, the steak is too cold or chewy, the soda is flat, or the bread is dry. The expectations for the texture, temperature, and touch play a vital role in our perceived quality of the food we eat. This has been shown in both research studies investigating behavior and scan studies investigating activation of the brain: A simple change in sensory information from the oral cavity can determine if the hedonic verdict is pure pleasure or disgust.

The complex array of sensory input from food can often give rise to misperception. Have you ever felt your mouth burn after eating a chili? Certain foods can give the illusion of temperature changes. The sensation of heat can arise from eating, e.g., chili (due to capsaicin), mustard (due to isothiocyanate), or pepper (due to piperin). This happens due to a chemical activation of heat temperature sensors—even though no temperature stimulation occurs. As these sensors under normal circumstances are only responding to heat, it will feel as if the food is extremely hot even though the food may even be cold. This specific temperature sensor that gives this sensation from chilis in humans is not activated by capsaicin in birds, which is why these animals take great pleasure in eating even the strongest chilis—presumably an ingenious trick of nature to ensure that the seed of the chili is being adequately spread by bird droppings. Meanwhile, a pinch of chili may be just what is needed to add a new dimension to a dish. This is an ingredient that should be used with a bit of caution, as the usage and perceived intensity differ drastically between cultures and individuals. For people without a sense of smell, the addition of chili may add complexity to the food but can in high concentrations, drown out the sensory input from the basic tastes and completely ruin the meal.

Temperature sensation is not just about heat. A cold sensation in the mouth can appear when consuming foods with menthol (e.g., peppermint or spearmint) or eucalyptol (e.g., common sage, sweet basil, bay leaves, or wormwood) due to the activation of cold temperature sensors. These sensations of heat/cold occur without any actual change of temperature in the mouth, even though the perception is quite real. In contrast, some sweeteners (e.g., xylitol or erythritol) can actually lower the temperature in the mouth, as the chemical transition from a solid to a liquid absorbs energy.

Texture and Flavor

Texture is defined as all the mechanical, geometrical, and surface attributes of a food perceptible by means of mechanical, tactile, and, where appropriate, visual and auditory receptors. It represents a dynamic tug-of-war between sensory and motor processes. Food texture starts with the eyes or possibly the nose, moves to the hands and then is delivered to the mouth. In the mouth, sensory signals provide guidance for the jaw and tongue movement activity, which changes the mouthfeel of the food. This constant feedback loop between sensory and motor is necessary to fracture the food into manageable pieces, reducing the risk of choking and readying it for digestion. Additionally, the fracturing of the food releases odors (along with their irritant qualities) internally into the nose and reveals taste to the tongue while reverberating the oral cavity, which is heard mainly through bone conductance.

Textures represent a vast array of descriptors. Texture terms are classified into three main classes of characteristics and subsequent primary or secondary properties (Table 4.1). Mechanical characteristics included texture properties that relate to applied stress (e.g., hardness, elasticity). Geometrical characteristics have properties related to structure and appearance (e.g., particle size and shape), while other

Table 4.1 Classification of textural characteristics by 3 main classes (mechanical, geometric, and other)

Primary	Secondary	Examples
Mechanical characteristics		
Hardness		Soft, firm, hard
Cohesiveness	Brittleness	Crumbly, crunchy, brittle
	Chewiness	Tender, chewy, tough
	Gumminess	Short, mealy, pasty, gummy
Viscosity		Thin, viscous
Elasticity		Plastic, elastic
Adhesiveness		Sticky, tacky, gooey
Geometric characteristics		
Particle size and shape		Gritty, grainy, course
Particle shape and orientation		Fibrous, cellular, crystalline
Other characteristics		
Moisture content		Dry, moist, wet, watery
Fat content	Oiliness	Oily
	Greasiness	Greasy

characteristics include mouthfeel properties that do not fall into the first two categories (e.g., moisture and fat content). As mentioned before, texture is not a static property of the food when eating, and it changes as you chew. In fact, this temporal aspect of texture is typically a desired quality in food, with combinations of textures that are in contrast to each other leading to increased enjoyment. For instance, a baguette would be very boring if it were soft all the way through the eating process. Luckily for us, it has lots of textures that contrast with each other. Your first bite hits the hard crust, which fractures, revealing a soft, airy interior, while your last bite is chewy and moist. Some descriptors are very unisensory, only being evaluated by touch, while many others pertain to different senses involved in their manifestations. For instance, listening is such an inherent part, along with touch, of several textures such as crispness that modifying the sound modifies the crispness. Marketers know this strong bond between sound and touch. Next time you buy a bag of chips, notice the noisiness of the packaging used; this acts as a priming effect, which increases your perception of the chips' crispness.

The perception of texture is intertwined with the other senses of flavor: the sense of taste can alter the perceived viscosity of food, where sweet stimuli increase the viscosity and sour stimuli decrease the perceived viscosity; if the viscosity is increased, the perceived smell and taste intensity decreases and the temperature can be perceived as lower. As such, the sense of touch must be regarded as an integrated part of flavor perception—and an important contributor to the construction of a sensory well-balanced meal. There are vast possibilities for variations in, e.g., texture, crunchiness, creaminess, and temperature when making a meal, many that do not require advanced tools for cooking or much extra time in the kitchen. There is a world of these possible flavor enhancers ready to explore and boost your food experience.

Handfeel, Mouthfeel, and Nosefeel

The sense of touch adds several pieces of the puzzle to the overall perception of food, often referred to as *mouthfeel,* but could further be broken down into its specific affected areas. When you eat, the sensory information from your fingers and hands holding the food or cutlery aids your evaluation of the food, both in regard to weight, surface, and texture (i.e., *handfeel*). Your lips, tongue, and the rest of the mouth participate in the further sensory analysis to help you be absolutely certain of what is in your mouth before you swallow. Both fingers, lips, and tongue have a disproportionately large brain area allocated for processing the information of touch, underlining the significance of this sense, and these feelings contribute to the texture of the food through mechanoreceptors. These feel types also contribute to chemesthesis (i.e., chemoreception) which activates through chemical stimulation rather than physical. Chemesthesis doesn't stop in the mouth but extends up the back of the throat into the nose. Touch sensation arising in the nose (or *nosefeel*) during the eating process has less to do with mechanoreceptors and more to do with chemoreceptors. Thus, it does not contribute to texture; it contributes to chemesthesis and thus flavor.

Summary

During eating, all senses can come into play and add to the complexity and pleasure of a meal. Likewise, the disruption of a single sensory modality can drown out or even ruin the overall experience of eating. Knowledge of all the senses can be a great aid when preparing and serving a meal. Furthermore, trying to break down and consider the contribution of each sense can be an eye-opening experience and a great tool for developing cooking skills.

Although eating is fundamental for survival, a disruption of pleasure from food and changes in habits and nutritional value may have severe consequences for both well-being and quality of life. The importance of harvesting the full potential of all senses and understanding what affects our eating behavior cannot be understated—neither in individuals with nor without a normal functioning sense of smell. The recipes of this book are made with special focus on a broad stimulation of several or all senses. With sufficient knowledge of how we sense and experience food, these recipes can potentially be improved and adapted according to individual preferences and sensory abilities. Therefore, we recommend revisiting this section or segments of the chapters in order to continuously improve both the ability to prepare and consume food with the highest possible level of enjoyment.

Additional Reading

Books

Shepherd, G. (2012). Neurogastronomy

This, H (2008). Molecular Gastronomy - Exploring the Science of Flavor

Piqueras-Fiszman, B & Spence, C (2016). Multisensory Flavor Perception: From Fundamental Neuroscience Through to the Marketplace.

Stafford, LD (2024). Smell, Taste, Eat: The Role of the Chemical Senses in Eating Behaviour.

Stuckey, B. (2012). *Taste: surprising stories and science about why food tastes good*. Simon and Schuster.

Articles

Matthen, M. (2015). The individuation of the senses.

Szczesniak, A. S. (2002). Texture is a sensory property. *Food quality and preference*, *13*(4), 215–225.

Pellegrino, R., & Luckett, C. R. (2020). Aversive textures and their role in food rejection. *Journal of texture studies*, *51*(5), 733–741.

Luckett, C. R., & Seo, H. S. (2015). Consumer attitudes toward texture and other food attributes. *Journal of Texture Studies*, *46*(1), 46–57.

Hyde, R. J., & Witherly, S. A. (1993). Dynamic contrast: A sensory contribution to palatability. *Appetite*, *21*(1), 1–16.

Witt, M. (2019). Anatomy and development of the human taste system. *Handbook of clinical neurology*, *164*, 147–171.

Chapter 5
Experience of Food

Contents

What Other Variables Contribute to Our Experience of Food

Our understanding and perception of food does not only rely on our senses but is also a product of a large array of processes in the brain and body. If you consider how differently you react to having a dry cracker when you are starving compared to one too many scoops of your favorite ice cream, the experienced pleasure can be remarkably reversed. Similarly, research studies have found that wine was rated better when participants were told it was an expensive wine. This chapter describes key variables that can change how we perceive and enjoy food.

Experience and Expectations

Largely based on prior experiences, we have a repertoire of sensory expectations: from sight at a distance to the smell before consumption is initiated to the mouthfeel and temperature when the food touches the teeth, tongue, and palate to the balance

A. W. Fjaeldstad et al., *Rediscovering Flavor*,
https://doi.org/10.1007/978-3-032-08056-1_5

between tastants to the crunchy sound during mastication to the aromas that enter the nasal cavity from the back of the throat.

All senses come into play during a meal, as we are evolutionarily optimized to detect potentially toxic elements—and to seek pleasure. If something doesn't match our sensory expectations, the sensory trait is analyzed, and the food is either rejected or swallowed. From the earliest stage, we depend on simple mechanisms to guide us in this decision. In newborns, the automatic response to sweet taste is licking the lips, searching for more sweet, nutritious pleasure. In contrast, sour and bitter foods cause an immediate withdrawal in newborns. This reflex is not limited to humans but can be found among most mammals (Fig. 5.1).

As we learn each time we eat, our repertoire becomes more advanced, and we use more senses to conduct this evaluation. These individual repertoires develop as a function of exposure and the intrinsic sensory and cognitive processes. No two individuals experience identical exposures through a lifetime. However, patterns of exposure can be identified, which can be related to geography, religion, and tradition—also referred to as culture. Even individuals with certain flavor-related professions and hobbies can differentiate themselves in terms of flavor subculture.

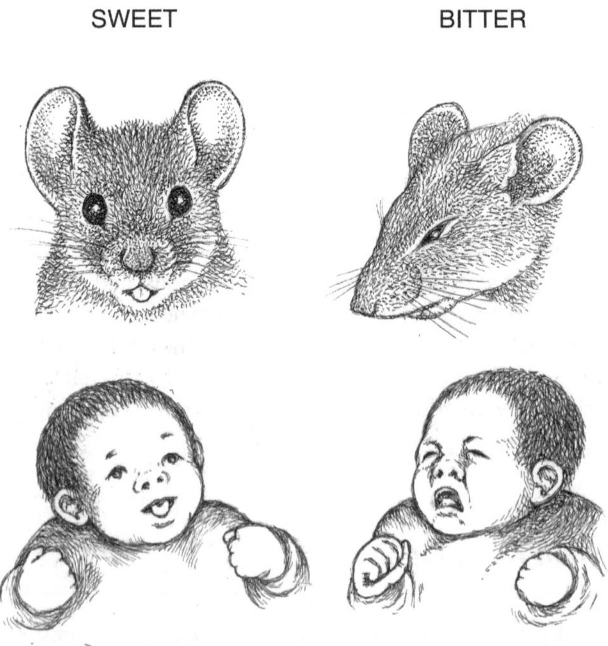

Fig. 5.1 Innate responses for taste evaluation. In both toddlers and other young mammals, there are shared innate responses for taste stimuli; sweet taste illicit tongue licking, while bitter foods illicit aversive gestures

Culture and Familiarity

One subculture that has been investigated extensively in flavor research is wine experts. It has been shown that the level of training affects the perception of wine, both in studies on behavior and in studies investigating activation of brain areas. Compared with wine novices, experts have larger activation in brain areas responsible for assessing texture and memory. In comparison, novices have greater activation in areas assessing pleasure/disgust. When asked "what do you think about the wine?", the responses make perfect sense: a common reply from the novice will be related to an evaluation of whether they like the wine or not, while the wine expert will take a more analytical approach and use color, texture, alcohol content, taste, and odor to determine the grape variety and origin of the wine. This example demonstrates two fundamentally different approaches to flavor perception.

Both expectations and experiences vary across cultures, which can have monumental effects on flavor and pleasure perception. In Denmark, salty licorice is among the most popular candies but gives little pleasure to individuals not accustomed to the taste. In many Asian cuisines, the presence of cartilage in meat dishes is considered a delicacy, while this is rarely served in Western dishes. Additionally, the combinations of spices and ingredients differ widely in Eastern and Western cuisines. This difference in cooking traditions gives rise to widely different perceptions of certain ingredients. As an example, vanilla is generally described as a sweet odor in individuals accustomed to the Western cuisines, as it is a common ingredient in desserts. However, in individuals accustomed to the Eastern cuisines, it may be perceived as spicy, as it is more commonly applied in such dishes there.

Not only does the qualitative perception change, but there are also examples of measurable quantitative differences in detection sensitivity between cultures. It has been found that non-industrialized populations are superior to industrialized populations when it comes to detecting certain odors in low concentrations. This difference may be driven by several factors, including better training due to a higher importance for procuring food, pollution impairments in industrialized countries, and environmental or genetic pressure. As such, culture can affect flavor perception on several levels, from better detection by the receptors in the nose to a more specialized processing of input in the brain.

The different frequencies of use of ingredients and availability of foods also affect the general familiarity with products and odors. In general, familiarity changes as a function of repeated exposure. There is a learning process, which especially seems to develop in late childhood and adolescence. This learning process is also central in developing an aversion for a specific food. It can occur after consuming a certain spoiled food or beverage that ended up causing nausea, vomiting, or sickness. In some instances, a bad food experience needs only to occur once before an aversion is generated that can last for years—or even a lifetime.

Individual Differences in Perception

Experience, expectations, and learning cannot explain all of the differences in flavor perception. Although training has been shown to especially alter how odors are processed in the brain, variation in perception is also driven by innate and genetic properties. Both on the sensory level and in the brain, there is a wide spectrum of different biological abilities for registering and processing sensory information.

Individual Differences on the Sensory Level

Almost no food gives the exact identical experience and reflections in two individuals; even simple tasks such as rating the sweeter fruit or agreeing on the origin of a smell can lead to diverging answers. From a sensory perspective, differences like these are quite expected, as our senses vary greatly, both in terms of sensitivity and perceived intensity.

The perceived texture of a given food can vary depending on the saliva flow and contents of enzymes, both of which are affected by genetics and state of the body (e.g., dehydration or infections). The taste varies both due to differences in the concentration needed to perceive the tastant (sensitivity) and the strength of the given taste (intensity), and in some individuals a tastant can elicit a completely different taste perception (e.g., some people taste a bitter tang in artificial sweeteners). This variation in perception does not become less significant when looking at the sense of smell. Here, the variation in sensory receptor activation between individuals for the same stimuli is on average 30%. For most smells, there is a difference in sensitivity, while other smells categorically sow dissension in pleasantness. A common example of this genetic variation is the polarized relation to cilantro—which for individuals who have a certain smell receptor expressed in the nose, causes a soap-like aroma, while individuals without this receptor perceive a discrete fresh, citrus flavor. Thus, many of these differences in both smell and taste have now been linked to different genetic versions of sensory receptors. As such, being particular about one's food can sometimes just be a matter of genetics.

Age can also play a role in how we perceive sensory input. In the same way as vision and hearing differ between individuals, so do the other senses. With age, many will require the use of glasses and hearing aids, but no such devices are yet available for improving the sensory tactile, taste, or smell stimulation—although efforts are underway in this direction.

Individual Differences in Processing Sensory Input in the Brain

An example of the differences in brain processing is synesthesia—a trait where the stimulation of one sense leads to an involuntary automatic perception of a second sense. The most common form of synesthesia is where numbers or letters have a color, but direct intertwined sensory perceptions are also found. In some people sounds are associated with colors, where environmental sounds (a car, a dog, or people talking) or musical sounds can trigger a firework of colors. Others experience auditory-tactile synesthesia, where specific words can create the illusion of touch. Finally, in word-taste (lexical-gustatory) synesthesia hearing certain words can trigger a sensation of a certain taste, which is experienced by 0.2% of the population. For these individuals, the sensation of taste is most often accompanied by a distinct sensation of both temperature and texture. It has been shown in neuroimaging studies of the brain that taste-related words actually activate the region of the brain normally reserved for taste stimuli. With the differences in processing of flavor in the brain and different peripheral sensitivity in mind, it is not surprising that there is a big difference in food preferences across individuals.

The Components of Pleasure: Wanting, Liking, and Learning

The perception of food and drinks is not static. Imagine how different a response can be when taking the first bite of a favorite dish, as compared with taking the last bite—especially if the portion was larger than the appetite. The yielded pleasure has a cyclic pattern, which can be subdivided into distinct behaviors and processes in the brain. In the appetitive phase, the wanting and craving for a given pleasure (food, drink, and other pleasures) gradually increases. In the brain, the neurotransmitter dopamine is primarily responsible for instigating this craving. At some point, this normally drives the individual to seek the wanted object (e.g., food or drink), after which ingestion can commence. This initiates the "liking phase," where another neurotransmitter (opioid) is released in the brain. This gives rise to the peak sensation of pleasure. However, after a certain amount of consumption, the pleasure sensation decreases, and the ingestion is terminated. Then, a phase of satiation follows. In this post-digestion phase, we are especially susceptible to reflections and learning, which can influence how the next pleasure cycle of wanting and liking is executed (Fig. 5.2).

Attention

Many factors can influence the wanting and liking phases. Apart from metabolic signaling, a common initiator of the pleasure cycle is attention, where an image or especially a smell can trigger attention toward a certain food. During consumption, attention is also crucial for the amount of pleasure gained from the meal. If attention is directed away from the food (e.g., by watching television or using a computer while eating), the perceived pleasure from the food is reduced and the time until the person once again initiates food-seeking behavior is reduced. In contrast, by enhancing the attention on the food (e.g., by consuming the meal in a social setting around

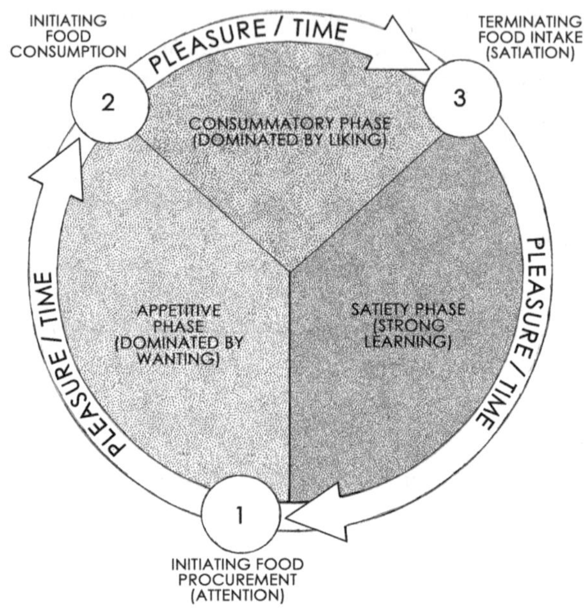

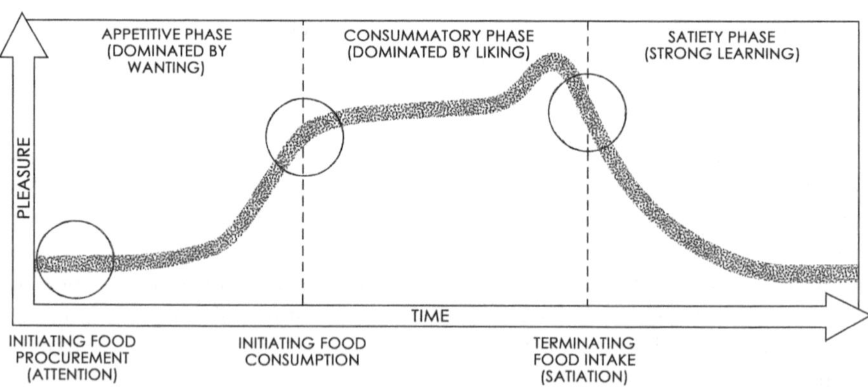

Fig. 5.2 The pleasure cycle. As with other pleasures, the pleasure of eating can be divided into separate phases, which can be described in a cyclic manner. The appetitive phase is dominated by wanting, where appetite and attention toward the given food or drink increase. This can be initiated by intrinsic cues, such as blood sugar levels, and by extrinsic cues, such as the smell or sight of food. When consumption is initiated, the linking phase represents the core of pleasure. As satiety sets in, both liking and wanting decrease until the cycle is at some point repeated. If a new course is introduced, the pleasure cycle may be initiated with wanting and liking, even if the stomach seemed full moments before. (Adapted from Fjaeldstad, A. et al. *Multisensory Flavor Perception* (eds. Piqueras-Fiszman, B. & Spence, C.) 211–234 (Elsevier Ltd., 2016))

a table and talking about the food), the sense of satiety lasts longer. This effect can also be accentuated when the food is stimulating more senses. As such, the old saying that "you eat with your eyes" can actually be taken quite literally.

Summary

Food is much more than just stimulating the senses. Some of the key factors in how we perceive food are experience and expectations, which have a huge influence on how, what, and when we eat. These cognitive processes combined with our abilities to perceive every single sense make every bite different— both for a given individual but especially between individuals. Trying new experiences and going out of the culinary comfort zone may be a rabbit hole into a whole new world of culinary pleasure—or disgust. We will never know the experiences of the paths not taken. Perhaps the greatest food experience lies ahead (Fig. 5.3).

Fig. 5.3 The social dining experience. When a meal is consumed as part of social interaction, both pleasure and the length of the satiety phase can be increased. As all senses are put into play, this effect can be even stronger. Eating is a multisensory experience. The sense of smell is stimulated as the food is being prepared and served at the table. The sense of hearing is stimulated by the sound of pouring the wine into the glass and the sound of breaking the crisp, freshly baked bread, which also stimulates the sense of touch in the fingers and mouth. The senses of smell, taste, and touch are stimulated throughout the meal. By sharing the food and talking about it, eating can truly be one of the strongest multisensory and pleasurable experiences

Additional Reading

Neurogastronomy How the Brain Creates Flavor and Why It Matters, Gordon M. Shepherd, Columbia University Press, 2011

Multisensory Flavour Perception. Spence, C. & Piqueras-Fiszman, B., eds. Elsevier, 2016

Smell, Taste, Eat: The Role of the Chemical Senses in Eating Behaviour. Palgrave Macmillan, 2024

Chapter 6
Appetite

Contents

Understanding Appetite and Hunger

Many of us have experienced that sudden, irresistible urge to eat when passing by a bakery. The scent of freshly baked bread triggers an immediate rush of saliva and a strong desire to indulge. Or perhaps the sight of an ice-cold drink on a hot day unleashes an instant craving, making everything else momentarily unimportant. These responses illustrate the power of sensory cues in driving our eating behaviors. But without these sensory triggers, how do we know when to eat?

This brings us to a crucial distinction: hunger and appetite are not the same. Hunger is a physiological state; a signal from the body that energy is running low and needs replenishment. This can manifest through physical sensations like a rumbling stomach or even the recognition of habitual mealtimes. Appetite, on the other hand, is psychological. It is the desire to eat, often sparked by sensory stimuli such as the smell, sight, or even the thought of food. While hunger ensures survival by prompting us to eat, appetite can push us toward food even when we're not physically hungry, or, conversely, suppress the urge to eat when we are.

At its core, appetite is governed by a complex interplay of hormones, neurotransmitters, and brain regions that regulate our desires and decisions. Our brain

A. W. Fjaeldstad et al., *Rediscovering Flavor*,
https://doi.org/10.1007/978-3-032-08056-1_6

constantly evaluates sensory information and past experiences, deciding which needs take priority at any given moment. This processing occurs in what is commonly known as the pleasure center of the brain, which integrates signals related to food, reward, and motivation. This was discussed in the previous chapter.

Under normal conditions, hunger increases the likelihood of stepping into the bakery when our blood sugar is low. However, appetite is not solely dictated by need; it is influenced by emotions, habits, and environmental factors. This explains why we may crave dessert after a full meal or lose interest in food when stressed, even if we haven't eaten for hours. The mismatch between appetite and actual energy needs plays a significant role in dietary challenges, contributing to both overeating and undereating.

The Role of Sensory Cues in Appetite

Sensory signals, particularly smell and sight, but also sound, are fundamental in shaping our appetite. The anticipation of food is often triggered before we even take a bite. The distant aroma of coffee, the sight of a perfectly plated dish, or the sizzle of food cooking can all stimulate the brain's reward system, generating a craving and even initiating physiological responses like increased saliva production.

Smell, in particular, has a unique role. Unlike sight or sound, which requires memory to predict how a food might taste, scent delivers a more immediate, visceral reaction. Remember, smell is involved before and after the food enters the mouth, which isn't true for sight. This is why the smell of baking bread can be so powerful. The smell doesn't just remind us of past experiences; it simulates them, activating brain regions associated with flavor and pleasure.

But what happens when these sensory cues are diminished, lost, or distorted? For individuals who have a sudden reduced sense of smell or none at all, appetite regulation shifts. Without the olfactory triggers that typically drive desire and expectation, food may seem less appealing, and eating habits may change. Some individuals report a decline in appetite, while others overeat, searching for satisfaction that taste and texture alone might not provide. Understanding this shift is crucial for adapting eating habits to maintain both nutrition and enjoyment.

Smell's Role in Driving Appetite

The role of smell in appetite is profound yet often subconscious. Unlike vision or hearing, which requires memory to interpret a food cue, smell delivers a direct line to the brain's memory and emotional centers. The scent of food can immediately evoke anticipation, stirring cravings that influence our desire to eat even before a meal is in sight.

At the core of this phenomenon is sensory-specific appetite, or our tendency to crave foods with particular nutritional qualities based on their aroma. For example, the smell of fresh bread may increase appetite not only for bread itself but also for other carbohydrate-rich foods like pasta or potatoes. This learned association helps us efficiently identify high-energy foods in our environment. Smell also plays a role in the cephalic phase response, where exposure to food cues triggers physiological changes such as saliva production, enzyme secretion, and hormonal shifts, priming the body for digestion.

However, our sense of smell is not easily deceived. While marketing often attempts to manipulate appetite through ambient food scents, research suggests that conscious awareness of an aroma does not reliably trigger appetite. Instead, subconscious exposure and contextual relevance, such as the scent of coffee in a cozy café, are more likely to activate the brain's appetite-regulating circuits.

Anticipation and Consumption

Smell informs our appetite in two key phases of eating: anticipation and consumption.

For anticipation, smell works at a distance, creating an expectation for food before it is even seen. The aroma of bread drifting from a bakery can make us crave carbohydrates, preparing the body for an incoming meal. However, once we approach the food, other sensory inputs may adjust or override the initial appetite. If we enter the bakery and see mold on the bread, visual cues override the appealing aroma, dampening appetite.

Once food is in the mouth, the sense of smell works in concert with taste and texture to shape our perception of flavor (as discussed in previous chapters). At this stage, external cues like sight and sound fade, and the proximal senses—taste, touch, and retronasal smell (aroma perceived from the back of the mouth)—take over. For example, an apple may appear fine on the outside, but biting into it might reveal an unpleasant texture and an off-putting odor, triggering a disgust response and causing rejection.

In both cases, the brain integrates sensory input with past experiences to determine whether food is desirable or repulsive. While many smells are universally perceived as appetizing or unpleasant, personal experiences and cultural background shape individual food preferences and aversions.

Smell, Salivation, and Conditioned Responses

The phrase "makes one's mouth water" is not just figurative as salivation is a key part of food anticipation, breaking down food particles and enhancing flavor. The Russian physiologist Ivan Pavlov famously demonstrated this with his dogs, who began salivating at the sound of a bell when it was repeatedly paired with food as mentioned in the section on hearing and flavor. Interestingly, he noted that smell alone was enough to trigger gastric secretions, reinforcing the idea that odor is one of the most potent sensory triggers for appetite.

This learned response works both ways. Just as pleasant food aromas can enhance appetite, negative experiences with food can lead to conditioned aversions. The Garcia effect (or Sauce-Béarnaise Syndrome) describes how a single unpleasant experience, such as food poisoning, can create a lasting aversion to the smell or taste of that food. A classic example is alcohol: after one bad night of overconsumption, the mere smell of it can trigger nausea for years, and this response is specific to the alcohol type consumed (e.g., tequila vs. rum).

This food conditioning begins forming long before our first conscious meal. Odor associations start in utero, as the diet of a pregnant mother subtly influences the baby's preferences. Newborns already exhibit instinctual attraction to sweet smells and avoidance of bitter ones—an evolutionary mechanism to seek nourishment and avoid potential toxins. Over time, these innate preferences are shaped by exposure and learning.

How Does Smell Loss Disrupt Appetite?

For many with anosmia (total smell loss), food becomes a functional necessity rather than an enjoyable experience. Without the ability to experience the full flavor of food, eating can feel like a chore rather than a source of pleasure. Some individuals report forgetting to eat altogether, only realizing they need food when physical symptoms, like shaking hands from low blood sugar, remind them to refuel. Others describe eating as a frustrating experience, leading to disinterest in social meals and avoidance of dining out, which can result in social anxiety and isolation.

However, anosmia doesn't affect everyone the same way. Some individuals lose interest in eating altogether, leading to reduced calorie intake and weight loss, while others overeat in an attempt to compensate for the lack of sensory satisfaction, often gravitating toward high-fat, high-sugar foods for their strong mouthfeel and residual taste components. This contrast suggests that appetite regulation in anosmia is highly individualized, likely influenced by psychological, hormonal, and metabolic factors.

Olfactory signals are necessary to drive appetite in a food-abundant environment, so what happens if we don't have access to this sensory information? The relationship between anosmia and food perception may seem straightforward, as smell plays a dominant role in flavor. Indeed, the loss of smell decreases the enjoyment of food—leading to less motivation to eat and changes to food-eating habits; however, its impact on appetite has not seen a lot of research, with most research studying behavioral outcomes of appetite disorders such as weight fluctuations. Many of the underlying physiological reasons have only been studied in non-primates, which doesn't always translate to humans.

Stress and Appetite

One of the most well-established factors influencing appetite is anxiety, and anosmia is frequently associated with increased stress and emotional distress. Anxiety-induced appetite changes follow a nonlinear relationship, often visualized as an inverted-U curve (known as the Yerkes-Dodson Law). Moderate anxiety may heighten alertness and even improve eating habits, but excessive stress can suppress appetite and food intake.

Stress triggers a cascade of brain-gut interactions that further complicate appetite regulation. The Hypothalamic-Pituitary-Adrenal (HPA) axis, a system that manages stress responses, plays a major role in regulating hunger hormones like ghrelin (which signals hunger) and leptin (which signals fullness). In some cases, chronic stress from anosmia may disrupt this system, leading to overeating ("stress eating") or undereating (appetite suppression), depending on the individual's coping mechanisms.

How Do We Restore Appetite?

The loss of smell can significantly alter appetite, but it does not necessarily lead to long-term dietary disruption. History provides valuable insights into this challenge. Ivan Pavlov, renowned not just for his research on conditioning but also for his surgical expertise, worked with patients experiencing digestive issues. He observed that some patients who were fed through tubes—bypassing the sensory experience of eating—gradually lost interest in food altogether. He described them as having "forgotten their stomachs," highlighting the importance of appetite beyond physical hunger. To improve digestion, he emphasized the need to restore appetite rather than simply providing nutrients.

As stated, changes in appetite following smell loss are highly individualized. A survey conducted in the 1980s by the Warwick Olfaction Research Group (WORG) found that while nearly half of smell-impaired individuals reported changes in their food preferences, about a quarter noticed no difference in their food experiences. And in many other studies, long-term anosmic individuals show no significant impact on appetite, weight, or dietary behavior. These findings offer hope as they suggest that compensatory strategies can be used to restore or maintain a balanced appetite. Over time, individuals can learn to recondition their appetite by shifting their focus to other sensory aspects of food, such as texture, taste, and temperature, or other aspects of the eating experience (e.g., dining out with a friend).

Restoring appetite is not about forcing oneself to consciously compensate in every meal. Instead, the goal is to naturally rewire associations so that food remains enjoyable and satisfying despite the loss of smell. This process, much like Pavlov's classical conditioning, relies on repeated exposure and new rewarding associations. The brain is remarkably adaptive, and while the initial loss of smell can feel overwhelming, many individuals find that through habitual exposure and positive reinforcement, their desire for food returns in new, meaningful ways.

Appetite is more than just a biological drive—it is shaped by experience, expectation, and emotion. While smell loss presents challenges, it does not have to mean the end of food enjoyment. Through a shift in focus to other pleasurable food attributes and other aspects of the meal, including cooking, appetite can be restored and even reimagined. We will explore these strategies in the coming section. By embracing new eating strategies and maintaining an open, curious mindset, individuals with anosmia can rebuild a fulfilling relationship with food, a relationship that extends beyond smell.

Additional Reading

Stafford, L. D. (Ed.). (2024). Smell, Taste, Eat: The Role of the Chemical Senses in Eating Behaviour. Palgrave Macmillan.

Harris, R., Harris, R. B., & Mattes, R. D. (2008). *Appetite and food intake: behavioral and physiological considerations*. CRC press.

Fine, L. G., & Riera, C. E. (2019). Sense of smell as the central driver of pavlovian appetite behavior in mammals. *Frontiers in physiology*, *10*, 1151.

Pearce, J. M., & Hall, G. (1980). A model for Pavlovian learning: variations in the effectiveness of conditioned but not of unconditioned stimuli. *Psychological review*, *87*(6), 532.

Pavlov, P. I. (1927). Conditioned reflexes: an investigation of the physiological activity of the cerebral cortex. *Annals of neurosciences*, *17*(3), 136.

Part III
Recovery Strategies

General Introduction

The sense of smell holds a remarkable secret: its capacity to regenerate. Unlike many other sensory systems, the olfactory system possesses the unique ability to renew itself after injury. While some systems in the body, such as the skin or liver, are known for their regenerative abilities, sensory systems often lack this capacity. For instance, hearing loss due to damage to the inner ear typically cannot repair itself, and vision loss from retinal damage is similarly permanent in many cases. In contrast, the olfactory system stands out as a rare exception among senses, and the potential for renewal highlights a powerful truth that healing is possible. While recovery may vary depending on individual circumstances and underlying causes, this healing characteristic offers hope and direction to those facing the challenges of olfactory loss.

This regenerative ability is not only unique but also surprisingly dynamic, even in cases where olfactory loss has persisted for years. Olfactory receptor neurons in the nasal cavity regenerate from basal cells, and studies in animal models suggest this process can take as little as 2–4 weeks. However, recovery involves more than just cellular renewal. These neurons must extend their axons, sometimes over several centimeters, to reconnect with the brain's olfactory bulb. Despite this distance, axons can grow at an average rate of 1 mm per day, meaning that the reconnection process may take just 1–3 months under optimal conditions.

While this timescale is strikingly short compared to other forms of nerve regeneration, such as in peripheral nerves or spinal cord injuries, it is important to remember that recovery is not always linear or predictable. For some individuals, particularly those who have gone years without smell, this regeneration process may still lead to improvement, as demonstrated by cases of delayed recovery. These stories of renewed olfactory function, even after long periods of loss, offer a cautious but hopeful reminder of the body's resilience.

Through an exploration of research in treatments, practical interventions, and patient experiences, we aim to equip you with tools to nurture your sense of smell. Recovery may take time, but with patience and persistence, progress may be within reach. Let's explore how to support your olfactory system on its journey back to health.

Chapter 7
Clinical Evaluation and Treatment

Contents

Diagnostics of Olfactory Loss

Before treatment for smell loss can begin, it is essential to understand its cause. A thorough diagnosis is the foundation for any effective treatment plan. The diagnostic process usually begins with a detailed medical history and symptom assessment, followed by a physical examination, specialized smell testing, and, in some cases, imaging of the nasal passages or brain.

The first and perhaps most important step is the clinical interview, during which the doctor will ask the patient about their symptoms in detail. Patients are usually asked to describe the nature of their smell problem—whether they have a reduced sense of smell, a complete loss of smell, or if they are experiencing distortions of smell. The doctor may also ask about how food tastes, since flavor changes are often noticed by patients. Essential details include when the problem began, whether the onset was sudden or gradual, and whether it followed an event such as a viral infection, head injury, or exposure to irritating chemicals. Past medical history, including any nasal or sinus diseases, surgeries, allergies, or chronic nasal congestion, is also reviewed. In addition, the doctor may inquire about occupational exposures,

A. W. Fjaeldstad et al., *Rediscovering Flavor*,
https://doi.org/10.1007/978-3-032-08056-1_7

smoking habits, medication use, and any family history of neurological diseases such as Parkinson's or Alzheimer's, both of which are known to affect the sense of smell.

Once the medical history is reviewed, the next step is a physical examination, often performed by an ear, nose, and throat (ENT) specialist. This typically includes a nasal endoscopy, in which a small camera is inserted into the nose to inspect the nasal passages and the olfactory cleft, the area high up in the nose where odor molecules are detected. This exam can reveal physical blockages, inflammation, or other conditions that might impair the ability to smell. This evaluation may also detect signs of infection or structural abnormalities.

In addition to the physical exam, smell testing plays a central role in diagnosis. These tests fall into three general categories: subjective reports, psychophysical testing, and electrophysiological or imaging-based tests (Fig. 7.1). Subjective reports involve self-assessment questionnaires in which patients rate their sense of smell. While helpful in capturing how a person feels about their condition, these reports are often unreliable. People may not notice mild or partial smell loss, or they may struggle to describe their symptoms accurately. For this reason, doctors often rely more heavily on psychophysical tests, which are standardized, more objective, and based on population size data to create medically meaningful distinctions between healthy and impaired. In these tests, odors are presented in controlled ways to evaluate the patient's ability to identify different smells, tell them apart, and detect very faint concentrations. These are often conducted using forced-choice procedures, where patients must choose between options rather than simply report whether they smell something or not. This approach helps minimize guesswork and improves accuracy. Two of the most commonly used tools in clinical settings are the University of Pennsylvania Smell Identification Test (UPSIT), which uses scratch-and-sniff cards with multiple-choice answers, and the Sniffin' Sticks, which are pen-like devices containing various odors used to test smell detection, identification, and discrimination.

In some complex cases, doctors may use more advanced techniques to measure the brain's response to odors. These include electrophysiological tests, such as EEG (electroencephalography), which records the brain's electrical activity during smell stimulation, and imaging techniques, such as functional MRI (fMRI), which shows areas of the brain that are activated by smells. These tests are not typically part of routine diagnostics and are usually limited to research centers or specialized clinics.

Finally, imaging studies such as CT scans or MRIs of the head may be used to visualize the structures involved in smell. These scans can reveal changes in the olfactory bulb or show signs of inflammation, injury, or shrinkage in key areas related to olfactory function. Imaging of the sinuses can also show mucosal swelling, mucus buildup, and nasal bone/cartilage abnormalities (e.g., deviated septum). Imaging is especially helpful when the cause of smell loss is unclear or when neurological disease or trauma is suspected.

In summary, diagnosing olfactory loss requires a careful combination of patient history, physical examination, smell testing, and occasionally imaging. Each step provides a piece of the puzzle, helping to identify the underlying cause and determine the most appropriate path forward for treatment.

WAYS TO TEST THE SENSE OF SMELL

SNIFF TESTS USING COMMON HOUSEHOLD ITEMS CAN YIELD A QUICK, INFORMAL ASSESSMENT.

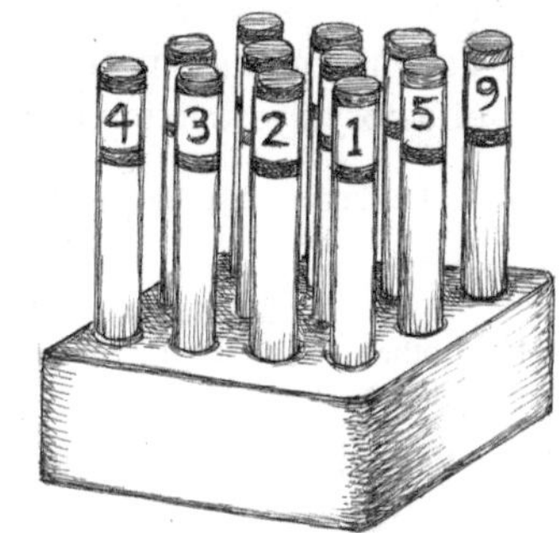

SNIFFIN'STICKS USED IN CLINICAL AND RESEARCH SETTINGS DETERMINE ODOR IDENTIFICATION, DISCRIMINATION AND A THRESHOLD.

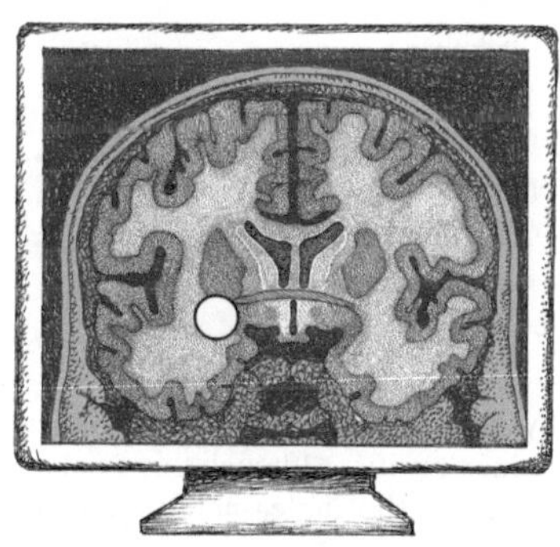

FUNCTIONAL MRI SCANS OR EEG BRAIN WAVE RECORDINGS SHOW HOW THE BRAIN RESPONDS TO DIFFERENT SMELLS.

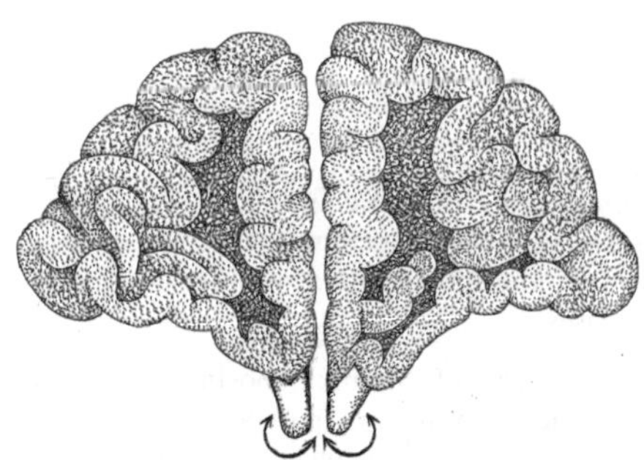

OLFACTORY BULB VOLUME MEASUREMENTS PROVIDE INFORMATION ABOUT FUNCTION AND PROGNOSIS.

Fig. 7.1 There are several different ways to test the sense of smell. While smelling common household items can give an indication of smell function, a validated test kit is required to accurately assess smell function and diagnose smell loss. Brain imaging and measurements of brain activity can add to the assessment of smell function in a clinical or research setting

Main Causes of Olfactory Loss

Olfactory loss can develop from various causes. In some individuals, the sense of smell gradually declines over time. In others, it can disappear suddenly after an illness or injury. While it may be tempting to think there is just one kind of smell disorder, the reality is more complex. Smell loss can come from problems in the nose, damage to the nerves that carry smell signals, or even changes in how the brain processes those signals. In this section, we explore the major known causes of olfactory loss, what is currently understood about how they affect the sense of smell, and what patients might expect in terms of recovery. An overview of the various causes of olfactory dysfunction is presented in the Table below.

Main causes of olfactory disorders with typical characteristics				
Cause	Onset	Prognosis	Parosmia present	Phantosmia present
COVID19 or other infections of the upper respiratory tract	Sudden	Improvement very often	+++	++
Chronic rhinosinusitis	Gradual	Very good treatment options	–	+[a]
Traumatic brain injury	Sudden	Improvement possible	+	++
Neurological diseases, such as Parkinson's disease, Alzheimer's disease, myasthenia gravis	Gradual	Improvement may be possible	+	+
Medicinal/toxic causes	Variable	Variable to good after discontinuation/ removal of the cause	+	+
Congenital anosmia		No therapy possible	–	–
Age	Gradual	Improvement possible	–	–
Other causes, e.g., B. iatrogenic damage (e.g., sinonasal and skull base surgery, laryngectomy), tumors, multiple systemic diseases	Variable	Improvement possible	+	+

[a]In many cases of Chronic rhinosinusitis, there can be a physiological explanation for the phantosmia, which is even more frequent in odontogenic chronic rhinosinusitis. Although this can be hard to distinguish from true phantosmia, this is not a phantom smell, as the unpleasant smell has a real intranasal source

One of the most common causes of smell loss is a viral infection, especially after a cold or flu. Many people became aware of this connection during the COVID-19 pandemic, where sudden loss of smell became a hallmark symptom. In most cases, the loss starts quickly, sometimes overnight, and for many people, recovery is possible. However, for others, it may take months or longer to regain a functional sense of smell. A common feature of this type of loss is parosmia, a distortion in which familiar smells begin to seem unpleasant or strange. Though distressing, parosmia is often a sign that the olfactory system is trying to heal, and with time, function may continue to improve.

Another frequent cause is chronic rhinosinusitis, an inflammatory condition affecting the nose and sinuses. This kind of smell loss usually develops gradually and fluctuates. Some people may not notice it's happening until the loss becomes significant. In these cases, inflammation itself, swelling, and nasal polyps can block odors from reaching the smell receptors high in the nasal cavity. In addition, long-term inflammation may damage the delicate smell tissues or even lead to changes in the brain regions responsible for processing smells. Fortunately, many people with this condition see improvement with treatments such as corticosteroids or surgery.

Unlike post-viral smell loss, parosmia is rare in this group, but phantom smells may emerge due to bacterial infection.

Head injuries are another important cause. Smell loss after a concussion or traumatic brain injury can occur suddenly and may result from damage to the nerve fibers that cross between the nose and the brain. These fibers can be torn or sheared during the injury. However, research also shows that damage may also occur deeper in the brain, not just at the nerve endings. This means that the injury might not be visible in a scan, but the person still experiences a loss of smell. Sometimes, the nasal passages are also affected, adding another layer of complexity. Recovery can happen, but it often depends on the severity of the injury and may take time. People with post-traumatic smell loss often report phantom smells or distortions, though these may lessen over time.

Neurological diseases, such as Parkinson's or Alzheimer's, are also linked to smell loss. In many cases, this is one of the earliest signs, even before memory issues or motor symptoms appear. The loss tends to develop gradually and is not always immediately apparent. For example, a person may start to enjoy food less or fail to notice common smells like smoke or perfume. In these conditions, the smell loss is related to changes in the brain's smell centers, including the olfactory bulb and deeper regions like the hippocampus. Because the underlying disease continues to progress, improvement in smell is generally not expected, and medications for the neurological condition do not usually help restore smell function.

Some people lose their sense of smell due to medications or exposure to toxins. Chemotherapy drugs, certain antibiotics, and even decongestant nasal sprays can affect the sense of smell. Industrial chemicals, heavy metals, and solvents are also known risks. The timing and severity of the smell loss can vary widely. In many cases, stopping the medication or avoiding further exposure to the toxin can lead to partial or complete recovery. However, this is not always guaranteed, especially if damage has occurred at the nerve or brain level.

Some people are born without a sense of smell. This is called congenital anosmia. It can occur as part of a broader genetic condition or appear on its own with no known cause. In these individuals, the olfactory bulb may be underdeveloped or completely absent. Because this is present from birth, most people with congenital anosmia have never experienced normal smell and only become aware of their condition when they realize they respond differently to odor cues than others do. Unfortunately, there are currently no known treatments that can restore the sense of smell in these cases.

Smell loss is also a part of the aging process. As people age, their ability to smell often declines, often without them noticing. The changes are usually slow and affect sensitivity to pleasant smells more than unpleasant ones. This might help explain why some older adults become less interested in food or fail to detect dangerous odors like smoke or gas. Aging-related changes affect not only the nose itself but also the olfactory nerves and brain structures. Additionally, previous injuries, infections, or toxin exposures may accumulate over time, leading to a more permanent form of smell loss.

In some cases, olfactory dysfunction may occur due to other medical conditions. Tumors in the nasal cavity or brain, surgeries involving the sinuses or skull base, radiation treatment, and certain systemic diseases like diabetes or thyroid disorders can interfere with smell. Psychiatric conditions and migraine have also been linked to changes in smell perception. The mechanisms vary and may involve blockages, nerve damage, or changes in brain chemistry. Recovery depends heavily on identifying and treating the underlying cause.

Finally, for a subset of people, no clear cause can be identified. This is referred to as idiopathic smell loss. These individuals often undergo thorough testing without a definitive diagnosis. Some of these cases may be due to unrecognized infections or early signs of a neurological disease. Although frustrating, this category highlights the complexity of the smell system and how much is still being discovered.

Understanding the cause of your smell loss can be the first step toward managing it. While some causes allow for recovery, others may require adaptation and support. In all cases, knowing the underlying reason helps inform expectations and treatment options, providing clarity in what can otherwise feel like a confusing and invisible health problem.

Treatment of Olfactory Loss

Treatments for smell loss depend on the underlying cause, and success can vary widely from one person to another. While some patients may recover naturally over time, others may need medical support. This section offers a summary of the most common treatment strategies currently available (Fig. 7.2). If you are experiencing smell loss, your healthcare provider will help determine which options are appropriate for your condition.

Medications and Other Drug Therapies

The most commonly used medications are corticosteroids. These are anti-inflammatory drugs, often used as nasal sprays, especially for patients with smell loss caused by chronic inflammation. When used for conditions like chronic rhinosinusitis, steroids—either in spray or oral form—can be helpful. However, for people whose smell loss is due to a virus or head injury, the benefits of steroid treatment are much less clear. Many studies have found little or no improvement in these cases, and steroids can have side effects, especially when taken by mouth.

Other types of nasal treatments have also been studied. One involves using a calcium-buffering spray, such as sodium citrate, which may help enhance the transmission of smell signals in the nose. Some early research has found positive results, especially in individuals who lost their sense of smell following a viral infection. However, further studies are needed to confirm the effectiveness of this treatment.

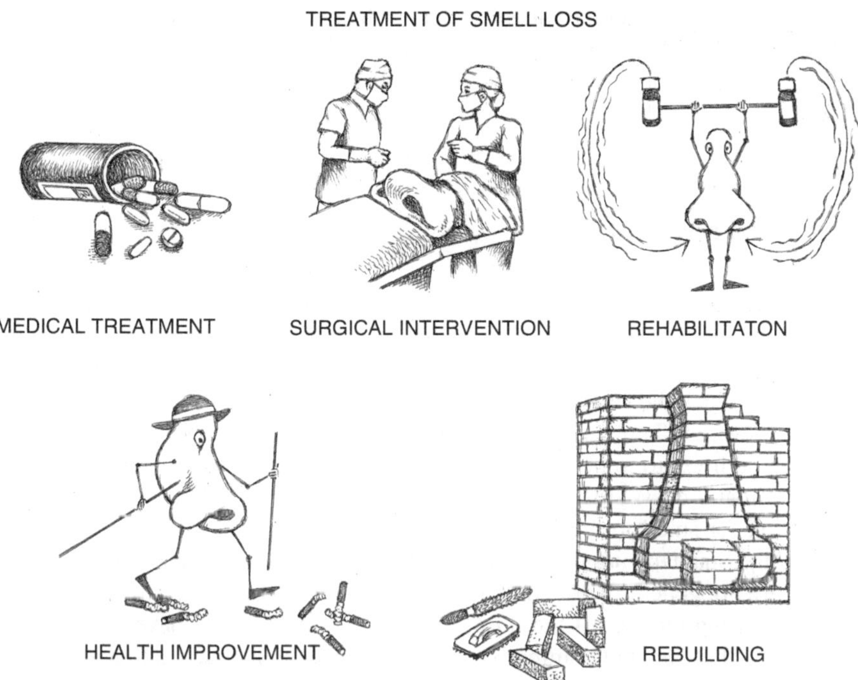

Fig. 7.2 Depending on the degree and cause of olfactory loss, there may be several treatment options, including medications, surgical interventions, and rehabilitation, such as olfactory training

Vitamin A has also been tested. This vitamin helps with the repair and growth of the tissue in the nose that detects smells. High doses administered orally showed promise in a small study, but larger trials using safer doses did not confirm the benefit. A few studies have investigated the direct application of vitamin A in the nose, yielding mixed results.

Another type of medication that has been studied includes phosphodiesterase inhibitors, drugs that influence cell signaling. These have been tested in both pill and nasal spray form, but so far, the evidence does not support using them regularly for smell loss.

Behavioral and Training Approaches

One of the most promising and low-risk treatments is called olfactory training. This involves sniffing a small set of different odors, such as rose, lemon, clove, and eucalyptus, twice a day for several months. The idea is to help the smell system recover through repeated stimulation. Multiple studies have shown that this type of training can improve smell, particularly in individuals who have lost their sense of smell due to a viral infection. While it does not work for everyone, the process is safe, inexpensive, and can be done at home. More details about how to do smell training are included in the next chapter.

Surgical Interventions

Surgery is sometimes recommended for people whose smell loss is caused by chronic inflammation, especially when nasal polyps are present. In these cases, surgery can remove obstructions and reduce inflammation, helping air and odors (and nasal sprays) reach the smell receptors more easily. Studies have shown that people with more severe inflammation and polyps often benefit the most from this kind of surgery. In addition to improved airflow, there may also be positive changes in the brain areas that process smells after surgery.

However, for smell loss caused by a virus or head trauma, surgery usually has little benefit. Standard nasal surgeries, such as septoplasty (to correct a deviated septum), are not generally effective in restoring smell. A few specialized surgeries, such as widening the part of the nose that odors travel through, have shown some potential, but these are not common, are not widely recommended, and need further studies for validation. In rare cases of severe phantom smells (phantosmia), surgical removal of the smell-sensing tissue has been attempted, but this is considered a risky and last-resort measure.

Biological and Regenerative Therapies

Some new therapies aim to help the smell system heal itself. One such treatment is platelet-rich plasma (PRP), which is made from the patient's own blood and contains healing factors. When applied to the nose, PRP may encourage tissue repair. Early studies in animals and small human trials have shown some improvement, particularly in individuals with post-viral smell loss; however, more research is needed before it can be considered a standard treatment.

Omega-3 fatty acids, commonly found in fish oil, have also been linked to better smell recovery. These fats appear to support the brain and nerve cells involved in the sense of smell. Some studies have shown that people who eat diets high in omega-3 fatty acids or take supplements may recover their sense of smell more effectively after illness or surgery.

Emerging Options and Technologies

Several innovative treatments are currently under development. One idea involves an "olfactory implant," which would function similarly to a cochlear implant used for hearing loss. This device would bypass damaged parts of the nose and directly stimulate the brain to produce a smell sensation. While this technology is still in early stages, it offers hope for the future restoration of smell in people with severe loss.

Scientists are also exploring the possibility of transplanting smell tissue or stem cells from the nose. These procedures have shown success in animal studies, with some animals regaining the ability to smell after treatment. However, these therapies are not yet ready for use in people and are still being researched in labs.

Other experimental options include herbal remedies, acupuncture, specific vitamins, and nerve block procedures. At this point, there is limited scientific evidence to support most of these, and they should only be considered in consultation with a healthcare provider.

Treatment of Qualitative Olfactory Disorders

Phantosmia: Phantosmia is often found in the course of olfactory loss following head trauma, but also in relation to other causes. The successful use of topiramate, verapamil, haloperidol, nortriptyline, valproate, phenytoin, and gabapentin has been described in single cases; however, informative studies are lacking. Topical application of saline solution to the olfactory mucosa may provide temporary relief. Transient improvement for up to 6 months has been observed after local anesthesia of the olfactory mucosa. Also, a significant decrease in postviral phantosmia has been reported after intranasal sodium citrate for 2 weeks. In cases of severe, prolonged phantosmia, surgical removal of the olfactory mucosa or the olfactory bulb has been used as a last resort in a few selected patients.

Parosmia: Because parosmia is typically associated with quantitative olfactory disorders, they are often treated together with quantitative olfactory disorders rather than separately. Parosmias are thought to resolve with the normalization of olfactory function.

While many treatments are still under investigation, there is growing hope for people affected by smell disorders. Some options, such as olfactory training and treating sinus disease, are already widely used and supported by evidence. Others, like implants and tissue regeneration, may become available in the future. If you are experiencing smell loss, a thorough evaluation with a qualified specialist can help determine the most appropriate treatment plan for you.

Additional Reading

Fokkens WJ, Lund VJ, Hopkins C, Hellings PW, Kern R, Reitsma S, et al. (2020) European Position Paper on Rhinosinusitis and Nasal Polyps 2020. Rhinology 58:1–464.

Doty RL. Treatments for smell and taste disorders: A critical review. Handb Clin Neurol 2019;164:455–479.

Holbrook EH, Puram SV, See RB, Tripp AG, Nair DG. Induction of smell through transethmoid electrical stimulation of the olfactory bulb. International forum of allergy & rhinology 2019;9:158–164.

Hummel T, Podlesek D. Clinical assessment of olfactory function. Chem Senses 2021;46:bjab053. https://doi.org/10.1093/chemse/bjab053

Patel ZM, Holbrook EH, Turner JH, Adappa ND, Albers MW, Altundag A, et al. International consensus statement on allergy and rhinology: Olfaction. International forum of allergy & rhinology 2022;12:327–680.

Whitcroft KL, Altundag A, Balungwe P, Boscolo-Rizzo P, Douglas R, Enecilla MLB, Fjaeldstad AW, Fornazieri MA, et al. (2023), Position paper on olfactory dysfunction: 2023. Rhinology. 61(33):1–108.

Rose H2020; 2022. https://rose-h2020.eu/. (accessed 07. May 2025).

Chapter 8
Smell Training

Contents

In 2009, olfactory training (OT) was formally introduced as a non-drug smell rehabilitation method, involving exposure to specific odors twice daily for 12 weeks. This study showed that approximately 28% of OT participants improved, while 6% without training experienced a decrease in function. But what exactly is training for the olfactory system? How does this training work to rehabilitate the nose? And how should it be for the most effective treatment?

What Is Olfactory Training (OT)?

Olfactory training (OT) is a non-medication approach to help people improve their sense of smell. It involves intentionally smelling a small set of distinct odors twice a day for several months. This method is used to support recovery in individuals who have lost their sense of smell due to infections, head injuries, or other causes. Olfactory training is a simple, inexpensive, and effective procedure, albeit somewhat boring. The typical routine includes four odors, such as rose, lemon, eucalyptus, and clove, that are smelled for about 30 seconds each, once in the morning and once at night. People are encouraged to continue the training for at least 12 weeks,

A. W. Fjaeldstad et al., *Rediscovering Flavor*,
https://doi.org/10.1007/978-3-032-08056-1_8

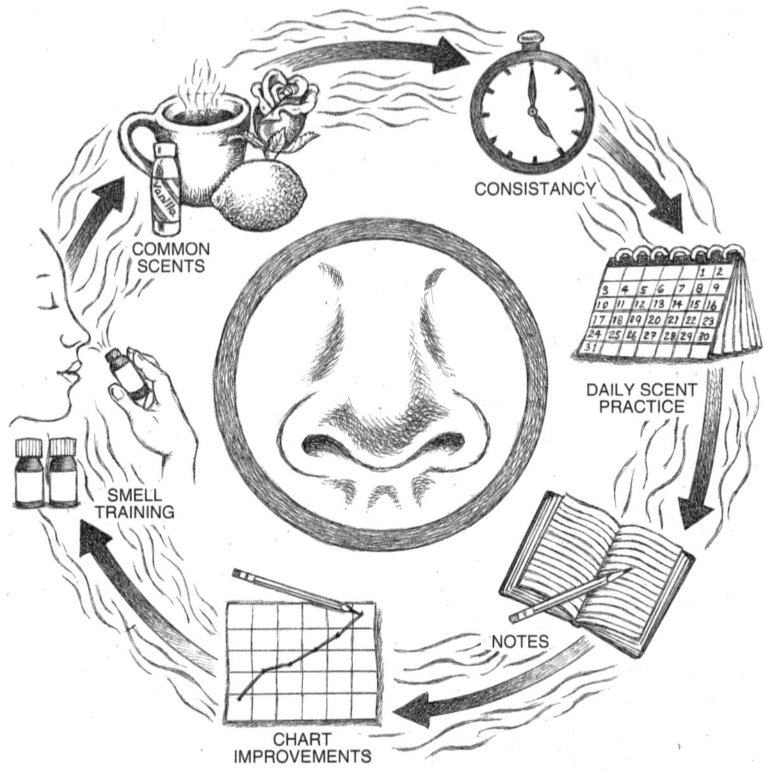

Fig. 8.1 Olfactory training

but ideally for up to 6–9 months. It is also encouraged for people to keep a smell journal to increase odor awareness (Fig. 8.1).

The training itself is simple, but the real challenge is consistency. Many people stop after a few weeks because they do not notice immediate changes, and the task itself can be less interactive the worse your ability to smell: Smelling jars without a sense of smell can seem pointless and even irritating. However, just as building muscle through regular physical exercise, retraining the sense of smell takes time and repetition. The odors used for training should be strong, consistent, and easy to identify. Commercially prepared essential oils or special odor kits are often more effective than household items like a freshly cut lemon, which may vary in strength and are less practical for long-term use. Over time, changing the training odors can help keep the practice engaging and might improve results. For example, menthol, thyme, tangerine, jasmine, green tea, bergamot, rosemary, and gardenia. Yet, increasing odors (to say eight odors) produced results comparable to those with fewer scents (such as the standard four). Additionally, diverse short sniffing exposures yielded similar benefits to longer sniffing sessions with fewer odors, indicating flexible approaches may enhance engagement.

Importantly, starting OT sooner rather than later may improve the chances of recovery. Individuals who start training within a year of losing their sense of smell tend to perform better than those who wait longer. This is a common effect with most treatments for an ailment. So use the QR code below to start training today, and feel free to make modifications that work with our schedule and liking.

How Does OT Work?

Olfactory training is not a recent invention. It has been known for some time that people can become VERY sensitive when they expose themselves to odors in their daily lives or professionally. For example, mothers can become extremely sensitive to the smell of their newborn babies—it is not only that they would know whether a given faint, delicate whiff of air contains the smell of a baby or not, but they can also tell the smell of their own baby apart from different babies. On a professional level, there are wine tasters, sommeliers, who develop an excellent sensitivity toward certain smells. The parts of their brains needed for the processing of smells also increase in volume, and they also respond more vigorously to odors. Comparative findings have been shown for professional perfumers and for women working in perfume retail outlets. Importantly, these increased abilities to smell improve the more these people work with odors.

Even in people with normal smell, repeated exposure to a specific odor can lead to increased sensitivity. For example, some individuals are unable to smell certain compounds, like androstenone, which is found in human sweat and some pork products. This condition is called a "specific anosmia," meaning the person is only unable to detect one particular smell. For androstenone, approximately ⅓ of the population has a very low sensitivity, a "specific anosmia," to it. However, when people with this kind of odor-blindness regularly sniff that odor over a few weeks, many begin to detect it. This shows that the olfactory system can adapt and improve when it is challenged repeatedly, which is the basic principle behind OT.

General Effectiveness of OT in Olfactory Loss

The effectiveness of OT varies depending on the cause of smell loss, the duration of the condition, and the consistency with which the person follows the training. Research shows that just a few weeks of training is often not enough to see results. Improvements are more likely after 12–16 weeks of regular training, and continuing the routine for up to 32 weeks or more can lead to even better outcomes. One study found that after 16 weeks, about 58% of participants had improved, and this increased to over 70% for those who continued for more than a year. However, it's important to note that "improvement" does not always mean a full return to normal smell function, but rather some degree of regained ability.

OT seems to work especially well for people who lost their sense of smell after a viral infection, such as COVID-19. Many studies now recommend OT for these patients, and some show that it helps people identify odors more accurately and may also reduce the unpleasant distortions of smell known as parosmia. Combining OT with certain medications or supplements may offer added benefit. For example, some people showed improved results when OT was paired with anti-inflammatory drugs, such as mometasone, or with natural compounds, like palmitoylethanolamide and luteolin.

For individuals who have lost their sense of smell after a head injury, results are more limited. Only about 1 in 4 people in these cases show meaningful improvement with OT. The severity of the injury and whether the person has other cognitive issues can affect outcomes. For people with unexplained or "idiopathic" smell loss, a few small studies suggest that OT can help, though the results are mixed. In cases where smell loss is due to chronic sinus inflammation, OT has not been extensively studied because other treatments, such as corticosteroids, are already effective. While OT after sino-nasal surgery can improve smell function, it still needs to be proven whether OT might provide added benefits in sinonasal disease.

OT also positively impacts olfactory performance in nonclinical populations, such as typical aging. Aging typically declines olfactory abilities, but OT may help slow this process, maintaining or improving functions in older adults. On the other side of the age spectrum, OT has been studied in children aged 9–14, showing improved olfactory performance, such as acuity and naming. In healthy adults, especially those who are regularly exposed to odors through their environment or hobbies, such as cooking and wine tasting, training has been shown to improve both odor detection and identification skills. In fact, increased awareness of odors correlates with increased olfactory performance, and simply increasing the conscious focus on odors may change their perception for the better.

In contrast to that, people working in odor-free environments, such as clean rooms, exhibit a lower olfactory acuity than people working in regular environments. Hence, environmental exposure to odors may also enhance olfactory functioning. To this point, engagement in cooking and attention to the odors involved (whether smelled or not) show a benefit for patients suffering from smell loss.

Possible Mechanisms of OT

Remember, when we smell something, tiny molecules from the air enter the nose and bind to special receptors located in the olfactory mucosa, a small patch of tissue high up in the nasal cavity. These receptors are found on sensory neurons, which send signals to the brain when they detect an odor. Damage to this system—whether from viruses, injury, or inflammation—can reduce the number or function of these neurons, making it harder to detect smells.

OT appears to help restore this system in several ways, and the exact mechanisms are still under investigation. In both healthy individuals and people with smell loss, OT has been shown to increase activity in the olfactory mucosa and to strengthen the signals that travel from the nose to the brain. Brain imaging studies show that individuals who engage in OT often have an increased volume in the olfactory bulb, a structure located at the front of the brain that receives signals from the nose. Other areas of the brain involved in smell, such as the orbitofrontal cortex and hippocampus, also show increased size or activity after OT.

In animal studies, OT leads to higher expression of certain genes in the nose and brain that are involved in cell survival and growth. These include genes that help neurons regenerate and adapt, suggesting that OT may not only strengthen existing pathways but also help the body build new ones. On the molecular level, OT may even encourage the expression of more olfactory receptors, enhancing the system's ability to detect a wider range of odors.

OT also appears to enhance the coordination between different brain regions. Functional MRI studies show that OT improves communication between regions responsible for processing smells, emotions, and even touch. This may explain why some people not only smell better after OT but also feel mentally sharper or more emotionally balanced (see next section).

Unconventional OT Tasks and Their Effects

Although further research is still needed, OT appears to support not just olfactory recovery but also aspects of cognitive and emotional health. These broader effects make OT an especially promising tool for aging populations and for individuals recovering from sensory loss related to illness or injury.

Several studies suggest that OT can improve cognitive functions like word recall and mental flexibility. For example, older adults who engaged in OT performed better on tasks requiring them to quickly generate words, compared to peers who did activities like Sudoku. Brain scans support these findings, showing increased activity in areas of the brain related to language and memory following OT. In fact, there is evidence that benefits from olfactory training may generalize to other sensory tasks, such as visual memory in adults. Similarly, when children train their sense of smell through structured exercises, they also seem to perform better on short-term memory tasks, including those involving visual information.

Some research has looked at emotional outcomes, such as changes in mood or symptoms of depression. While OT does not appear to affect mood in healthy individuals or in people with diagnosed depression, there is some evidence that it may reduce depressive symptoms in older adults with mild or subclinical depression. These findings suggest that OT may offer modest emotional benefits, possibly due to increased engagement, improved sensory stimulation, or subtle changes in brain function.

Summary

Olfactory training is a simple, safe, low-cost, and effective method for improving the sense of smell in people with various causes of olfactory loss. It works by stimulating both the nose and the brain, encouraging recovery through regular exposure to odors. While the exact molecular processes are still being explored, research indicates that OT is particularly beneficial for post-infectious olfactory loss and may also offer benefits for individuals with other types of sensory loss. In addition to improving smell, OT has also been linked to gains in memory, language, and emotional well-being, suggesting that its effects may reach beyond the olfactory system. These improvements appear to be stable over time, making OT a valuable tool for both rehabilitation and healthy aging.

Further Reading

Review

Pieniak M, Oleszkiewicz A, Avaro V, Calegari F, Hummel T (2022) Olfactory training - Thirteen years of research reviewed. Neurosci Biobehav Rev. 2022 Oct;141:104853.

Specific Research Articles

Hummel, T., Rissom, K., Reden, J., Hähner, A., Weidenbecher, M., & Hüttenbrink, K. B. (2009). Effects of olfactory training in patients with olfactory loss. *The Laryngoscope*, *119*(3), 496–499.

Damm, M., Pikart, L. K., Reimann, H., Burkert, S., Göktas, Ö., Haxel, B., et al. (2014). Olfactory training is helpful in postinfectious olfactory loss: a randomized, controlled, multicenter study. *The Laryngoscope*, *124*(4), 826–831.

Patel, Z. M., Wise, S. K., & DelGaudio, J. M. (2017). Randomized controlled trial demonstrating cost-effective method of olfactory training in clinical practice: essential oils at uncontrolled concentration. *Laryngoscope investigative otolaryngology*, *2*(2), 53–56.

Olofsson, J. K., Ekström, I., Lindström, J., Syrjänen, E., Stigsdotter-Neely, A., Nyberg, L., et al. (2020). Smell-based memory training: Evidence of olfactory learning and transfer to the visual domain. *Chemical Senses*, *45*(7), 593–600.

Fjaeldstad, A. W. Culinary Cure? Improved Subjective and Measured Sense of Smell after a Cooking Rehabilitation Program. Int. J. Gastron. Food Sci. 101121 (2025) https://doi.org/10.1016/j.ijgfs.2025.101121.

Oleszkiewicz, A., Heyne, L., Sienkiewicz-Oleszkiewicz, B., Cuevas, M., Haehner, A., & Hummel, T. (2021). Odours count: human olfactory ecology appears to be helpful in the improvement of the sense of smell. *Scientific Reports*, *11*(1), 16888.

Chapter 9
Cooking and Compensation Strategies

Contents

Getting the Most Out of Food

Food can be subdivided into cooked and uncooked (raw) food. Some uncooked food, such as broccoli, contains biologically available nutrients and is ready to consume, freshly picked from the stalk. Other uncooked foods, such as meat, require the addition of heat or other forms of energy to make nutrients available and easily consumed through chewing. Historians have often credited the advent of cooking as a primary reason humans branched off from the Great Ape family and started developing a large neocortex. Here, cooking allowed us to waste less time chewing complex proteins and more time reaping their benefits. Cooking has the additional

A. W. Fjaeldstad et al., *Rediscovering Flavor*,
https://doi.org/10.1007/978-3-032-08056-1_9

benefit of making food safe for consumption by lowering the number of pathogens, thus decreasing the risk of infection.

For someone with smell loss, understanding the basic science behind cooking can increase your skill and confidence in the kitchen. This knowledge, along with the aid of your other senses, can help you determine when food is cooked correctly, or "finished." Like gravity, the physics of food obeys certain laws leading to expected outcomes. For example, simply knowing how to measure a physical unit such as temperature helps guarantee a consistently cooked steak, whether that be medium rare or well done. Just as leaning into the social aspect of eating can be rewarding, using science to create new sensations can be stimulating and lead to an even higher appreciation of the cooking process. It also doesn't hurt your confidence to impress your friends and family with the new knowledge.

Cooking can be a never-ending hobby. Consider the wide range of cuisines and the regional or even hyper-local differences within them. We're talking millions of recipes! Now consider learning *why* certain aspects of these recipes work and *why* others don't. Not just the simple answer of "because the mayonnaise will break," but a more in-depth understanding of oil and water emulsions. Understanding the "why" behind cooking isn't just about knowing "what" works—it's about the science behind it. For example, learning *why* pesto browns tells you *what* to do to avoid it and even *what* to do to bolster the greenness of it. Even replacing single ingredients within recipes may transform the recipes into new ones. The process of fermentation, for example, completely changes the taste and the texture of a cucumber, adding a sour and prickly sensation. Now imagine how you can apply this new knowledge to modify existing or create new recipes. This is truly the joy of cooking, and the number of sensory neurons functioning in your nose does not limit that joy. Understanding and applying the science in the kitchen brings a new appreciation to food preparation and allows you to alter tastes, mouthfeel, and appearances underrepresented in standard recipes.

In this chapter, we will provide some of this knowledge to help you succeed in the kitchen and have fun while you're in there. We emphasize "some" as we're covering topics from an entire discipline known as Food Science, and there are far too many topics to discuss them all in a single chapter. So, think of these pages as a primer to a completely new outlook on cooking, and know there are many adventures that await in texts that are more comprehensive. We'll help guide you in the right direction and provide a few resources for in-depth investigation in a *further reading* section at the end of this chapter.

What Is Cooking?

Our universe is made of just two things: matter and energy. Cooking is the transformation of a food, a source of matter, with energy. In cooking, like most human activities, matter and energy do not vanish; however, both can change form. Water can change from liquid to gas with energy (such as heat), yet the water has not changed its mass (or weight when considering gravity). Thus, cooking happens as a chain of reactions. Recall the grade school activity of combining baking soda with vinegar to create a lava-like foam of bubbles—usually bellowing out of a paper mâché volcano. This is a chemical reaction between a base (the baking soda) and an acid (the vinegar). You can write this as a simple formula in common and scientific language:

Common
Vinegar + Baking Soda → Carbon Dioxide + Water + Sodium Acetate

Scientific
CH_3COOH (Acetic acid) + $NaHCO_3$ (Sodium bicarbonate) → CO_2 (Carbon dioxide) + H_2O (Water) + $NaC_2H_3O_2$ (Sodium acetate)

In this reaction, acetic acid (the main component of vinegar) reacts with sodium bicarbonate (baking soda) to form sodium acetate, carbon dioxide, and water. The carbon dioxide gas produced leads to the bubbling and fizzing, which acts as the faux lava coming out of the faux volcano. The mix of these two substances is common in the kitchen, having a practical effect used in cooking, especially in leavening baked goods.

You can even speed up or slow down reactions with a catalyst or some additional agent. Let's refer to these agents of change collectively as "reaction modifiers." Reaction modifiers like heat and pH can affect the rate, direction, or outcome of chemical reactions. When we consider heat's role in the reaction above, we need to consider that all reactions need energy to happen, and heat can speed up the reaction by providing the energy that helps overcome some activation energy barrier. In other words, increasing temperature speeds up the reaction by providing more kinetic energy to the reactants, leading to quicker formation of products.

In cooking, you have several reactions happening all the time so it's easier to think of a more general formula of the process. However, the same idea of a chemical reaction can apply—you have some raw ingredients that undergo some transformation from uncooked to cooked, often under the influence of a reaction modifier (like heat). This process changes the food's texture, taste, and appearance. Using notions derived for basic chemistry we can extrapolate the cooking process into a simple, general formula:

$$\text{Raw ingredients} \rightarrow \text{Cooked Food}$$

Frying an Egg: Common Chemical Reactions in Cooking

Let's now consider the raw ingredient of eggs undergoing the popular cooking reaction modifier, heat. Applying heat to eggs does not create a single chemical reaction but a series of reactions that involve the denaturation of proteins, loss of water, and the Maillard reaction, among others. This drastically changes the color, shape, and

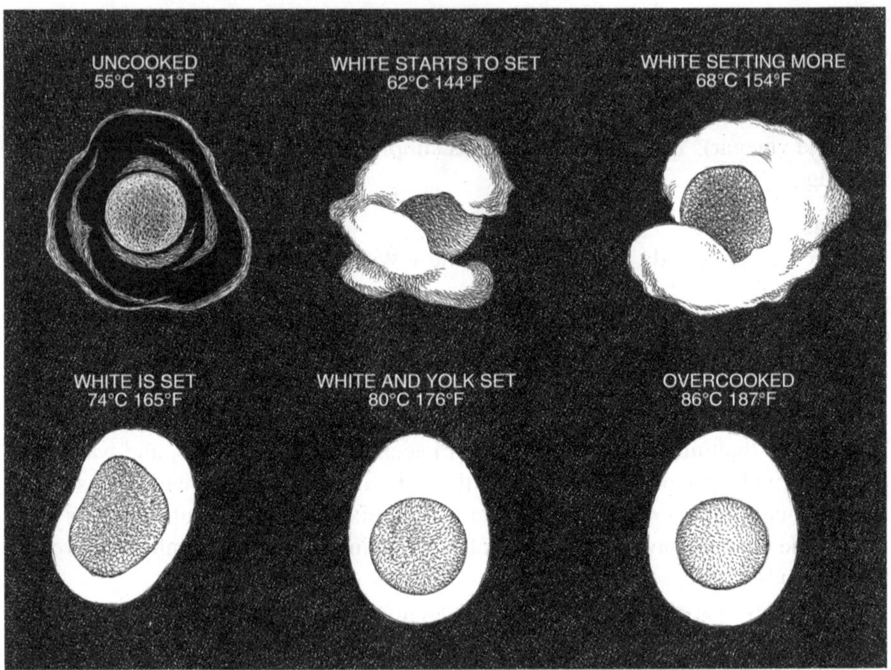

Fig. 9.1 Transformation of egg from raw ingredient to cooked food

texture of the egg, moving the raw egg ingredient from uncooked to cooked (see Fig. 9.1).

For the sake of creating a simple and educational formula, let's focus on some key processes that are involved in this transformation:

Protein Coagulation

Initial Stage As the temperature of an egg rises from applied heat, the proteins begin to denature, meaning they unfold from their natural, coiled structure. This process starts at temperatures as low as 63 °C (145 °F) for egg whites and slightly higher for egg yolks.

Coagulation Following denaturation, the unfolded proteins start to bond with each other, forming a gel-like network. This network traps water and forms the solid structure characteristic of cooked eggs. The coagulation process for egg whites occurs around 65 °C to 70 °C (149 °F to 158 °F), and for egg yolks, it occurs around 70 °C to 73 °C (158 °F to 163 °F) (Fig. 9.2).

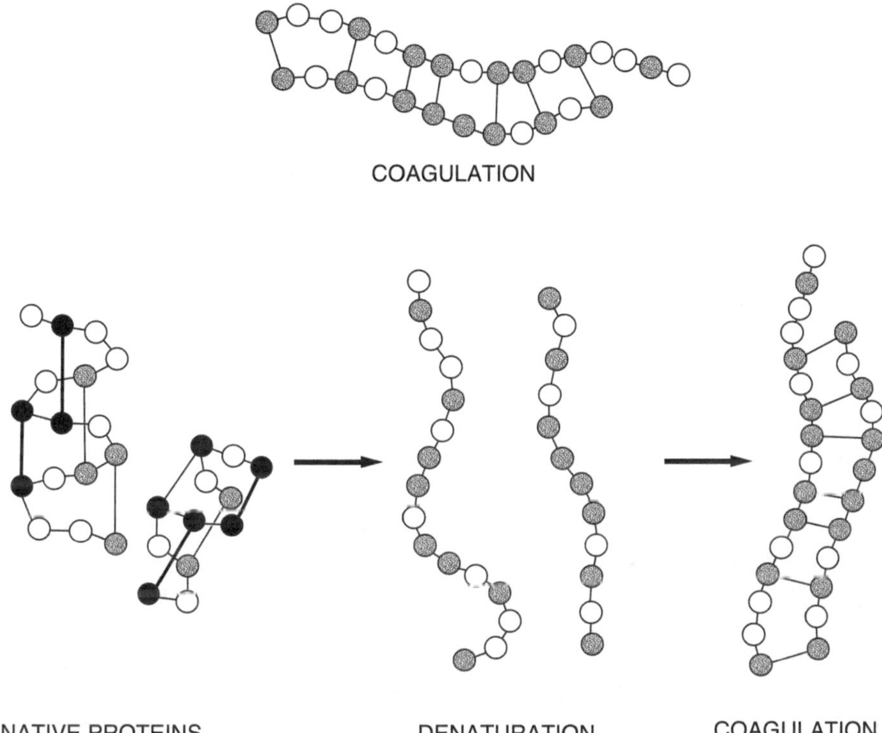

Fig. 9.2 Changes in proteins during heating

Moisture Loss

As the temperature increases, moisture loss will occur primarily through two mechanisms: the reformation and tightening of protein structures in coagulation, which squeeze out water, and evaporation at temperatures higher than the boiling point of water (100 °C/212 °F). Thus, moisture loss will occur at the surface first, but prolonged heating will release additional water from protein coagulation. The trapped water leads to a juicy/tender texture, but without it, an egg will have rubbery or tough textures.

Maillard Reaction

This reaction involving sugars and proteins is crucial for developing complex flavors, tastes (umami), and browning; it starts at higher temperatures, around the boiling point of water, and continues as the temperature rises, peaking at 160 °C (320 °F). Think of the toasting of bread or the crust of a pan-seared steak. While less pronounced in eggs than in meat due to the lower content of sugars, the Maillard reaction still occurs, especially in egg dishes cooked at higher temperatures that have undergone moisture loss, such as omelets or fried eggs.

Knowing these processes, a simplified cooking formula could look something like this:

$$\text{Uncooked Egg} + \text{Heat}\left(\text{Reaction Modifier}\right) \rightarrow \text{Denatured Proteins} \\ + \text{Water Loss} + \text{Maillard Products}$$

In this egg example, water, proteins, and sugars are integral to the shape and texture of a food, and reaction modifiers will alter their stable, uncooked state. The most common reaction modifier in cooking is heat.

Heat

The longer you apply a heat, the higher the temperature of the ingredient and the more cooked the raw ingredients. Heat is the most common form of energy in cooking used to change the molecular structure of food. The science of energy is *thermodynamics*. For cooking, heat can come in many forms, from radiation (via microwaves) to chemical heat (via combustion from propane gas). In thermodynamics, the second law is very important for cooking, which states that heat naturally flows from hot to cold. In cooking an egg, the hot pan transfers heat conductively to the surface of the egg and moves inwards to its colder center with time. This law is why you can cook a soft-boiled egg and medium-raw steak, and why it's important to measure the center of a food to determine doneness. However, with enough heat and time, there will be no difference in temperature at the surface and the center. This point of cooking is also discussed in thermodynamics as the *equilibrium state* of the food. The rate to heat food is dependent on the transfer of heat and the food (or matter) it's heating.

Each food has an estimated heat capacity (C_p), often reported as a unit relative to water ($C_p = 4.18$ J/g °C), which determines the amount of heat needed to increase its temperature. Importantly, a high heat capacity means a food that takes longer to become hot will also take longer to cool down. Most foods contain a lot of water; therefore, their heat capacity is high, similar to water (C_p = 1) (Fig. 9.3). Eggs,

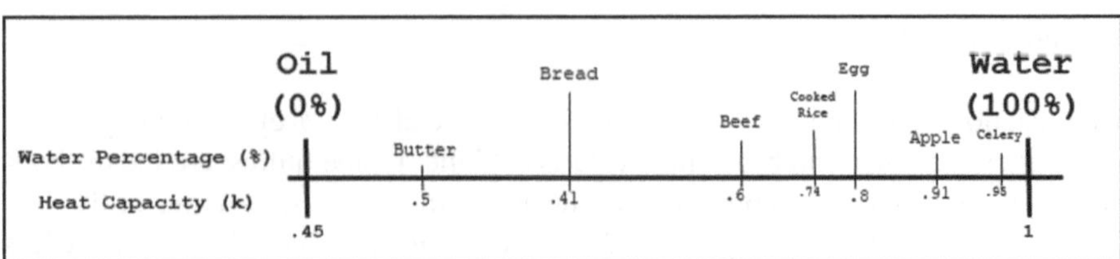

Fig. 9.3 Water content and heat capacities of common foods

which are approximately 74% water, have a heat capacity close to that of water ($C_p \approx 0.91$). Conversely, fats and oils have heat capacities about half that of water (~0.5). Meats, containing both fat and water, have a heat capacity that falls between oil and water. Importantly, the heat capacity is influenced by water content *and composition*, so bread has a lower heat capacity than oil due to its *starch, protein, and air pockets*.

The high-water content of foods also makes most food thermal insulators. Heat, therefore, travels slowly from the outside surface of food to the inside middle. Water has a lower thermal conductivity ($k = 0.0056$) making it a perfect medium for slowly cooking something like a poached egg. Conversely, thermal conductors (k = 100–400), such as metals in pots, are used for transferring heat to the food quickly. Cooking vessels made of different materials will transfer and hold heat differently. Mixed copper pans are great for quick sautés, while ceramic pots might be better for long braises. Considering a typical metal pan, a useful "rule of thumb" is that cooking time is roughly proportional to the square of the thickness of food. This rule holds for conductive cooking (direct contact of two solid objects like a pan and egg), but there are other means of heat transfer. Thermal convection deals with fluid contact (from a liquid or gas), which happens when poaching an egg, while thermal radiation occurs when you microwave an egg (as electromagnetic energy waves are absorbed). However, all types of heat follow the second law of thermodynamics, transferring their heat from hot to cold. Most cooking processes involve a combination of thermal transfer types, and cooking devices are variable across kitchens, making it hard to quantify exact cooking times. Recipes give estimates on time and temperatures, but experience and a good thermometer will give you the best estimate. Remember: Heat transforms food. And with these transformations come new shapes, colors, sounds, and textures. Learn to cook by these sensory cues, and you'll become a better cook who doesn't even need a thermometer.

Water

Water is a magical chemical compound that is simple. Water molecules are attracted to each other and other important charged molecules in food (such as salt) due to shape and chemical charge (Fig. 9.4). Water is everywhere and is essential to all living objects. We drink, cook, and clean every day with water. Water is in plant and animal cells and plays a major role in foods' texture when eaten raw or cooked. A food with a high percentage of water is referred to as "juicy," and foods with a low water percentage are referred to as "dry." Remove enough water, and the food will be preserved for several years until you rehydrate it. Sun drying has been a common method of preservation since ancient times. With time, adding heat can turn liquid water into a gas, and removing heat can make it a solid. Liquid water has a large range of temperatures (0 to 100 °C) and most foods have a final cooking temperature within this range. Consider an egg that cooks at 62 °C or the safe-to-eat internal temperature of pork [63 °C (145 °F)]. For this reason, a common modernist cooking

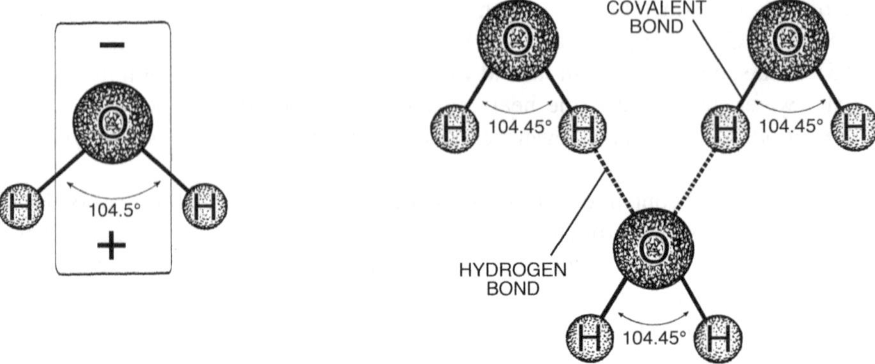

Fig. 9.4 Illustration of a water molecule

method involves cooking foods in a controlled water bath set at the final temperature desired for the food (called sous vide). Importantly, pure liquid water will never go above the temperature of boiling at atmospheric pressure. Unique to substances, liquid water is denser than its solid form, as the molecules expand under cold conditions to form a near-perfect cubic arrangement. This leads to ice floating on top of water as you see with icebergs. Water vapor (or steam) is above 100 °C thus can cook foods quicker than boiled water. Liquid water can be raised to even higher temperatures with pressure (using kitchen tools like a pressure cooker). If you wanted the quickest cooking method for a hard-boiled egg, a pressure cooker would be the fastest, next to steam and then the traditional method of boiling.

Understanding how to control water within a food is vital to controlling its texture. To reach higher cooking temperatures needed to create browning and crispiness, such as for fried chicken or an egg, water needs to be removed. It's the difference between a tough steak and a crispy, juicy steak. Pressure and the gas-water phase are one way to add more heat than boiling water, but a more common cooking technique is to cook in fat. Fat can reach much higher temperatures than water and can impart flavor to the food being cooked. Cooking in fats is easy to control, as they have a lower heat capacity than water (see section "Heat")—meaning they can gain and lose heat quickly. There are many fats with different *smoke points,* or a maximum cooking temperature of the oil until it decomposes and produces off flavors. And different fats have different flavors, from neutral (peanut oil) to pumpkin (pumpkin seed oil) to porky (lard). Peanut oil is ideal for frying, as it's neutral in taste and has a high smoke point of 400 °F/200 °C. Increasing the outside temperature of foods rapidly with fats will evaporate water reaching the surface. The dryer surface will further increase in temperature, start to brown, and create textures like "crispy." However, heating a food for too long will eventually dry it out.

In summary, heat is a common source of modifying the chemical structure of foods, and most foods are composed of water. At lower temperatures, protein coagulates as water increases in temperature, and at higher temperatures, water is released and then evaporates, leading to temperatures above boiling point that modify sugars. These higher temperatures are what lead to browning and some intense flavors formed from complex interactions between sugars and proteins. Detailed information about interactions of protein and sugars with heat is beyond the scope of this book, but two important reactions at these higher temperatures are important to drastically increase flavor.

Adding Intense Flavor at High Temperatures

Believe it or not, when you cook food at high temperatures, you're creating many more flavor molecules than in the same food uncooked. Sugar and proteins, along with fat and water, are the main components that shape the texture and flavor of food. Water contributes to juiciness, and fat adds tenderness; these famously don't mix because fat doesn't have a charge. Sugars and proteins, however, dissolve in water and play an additional, exciting role. They provide structure to food, and their superpower lies in their ability to break apart and recombine in entirely new ways—transforming themselves and the flavors of the food in the process.

Caramelization and the *Maillard reaction* are two processes that create these transformed flavor molecules. The key difference is that caramelization involves only sugars, while the Maillard reaction requires both sugars and proteins. The Maillard reaction occurs when proteins and sugars in food are exposed to high heat, typically above 300 °F (150 °C). This reaction creates a cascade of complex, savory, nutty, and roasted flavors. Think of the golden crust on bread, the sheared edges of a steak, or the deep color of roasted coffee beans—these are all thanks to Maillard reactions. Caramelization, on the other hand, happens when sugars are heated to even higher temperatures, typically above 320 °F (160 °C). As sugars break down and rearrange, they develop rich, sweet, and slightly bitter flavors. Caramelized onions, toffee, or the golden top of crème brûlée are delicious examples of this process.

Diffusion and Osmosis

We earlier discussed the second law of thermodynamics about temperatures moving from high to low. A similar principle, but more general, refers to the movement of something from high concentration to low. For most seasonings, such as salts, this is referred to as *diffusion*, but a special case of this for water is *osmosis*. These chemical processes are how a cell functions to bring in or out nutrients (and water) and how a brain neuron spikes to send information. They describe nature's tendency to seek an equilibrium state. In cooking, these processes are necessary for an important concept in cooking: seasoning foods from within so each bite is flavorful. Foods undergoing diffusion and osmosis not only change in flavor but also texture. Brining a chicken allows salt to penetrate the whole chicken while also adding "juiciness" from water moving inside the meat. In fact, often both processes are working in parallel—spice and water trying to find a new equilibrium state. Diffusion is mainly affected by three factors: water, temperature, and time. As with many things in cooking, diffusion is sped up by heat and the amount of water in the food. Higher water percentage and even cooking in water encourage faster diffusion. For instance, a quick method to penetrate flavor into a food is cooking it in water (as commonly done with braising meats). This is also a reason to heavily salt blanching and poaching water for pasta or vegetables. Salts, sugars, and acids are common agents of flavor that take advantage of diffusion. Coming back to the simple egg, we can see these chemical processes working in tandem when making salt-cured egg yolks (Example Exercise—see box). Osmosis explains how water is drawn out of the egg to create a harder texture, while diffusion explains the saltiness within the egg.

Exercise: Osmosis and Diffusion from Curing Eggs with Salt

Ingredients

12 eggs

A lot of salt

Method

In a non-corrosive pan, like a baking sheet or cast-iron skillet, make ½ inch (1.5 cm) layer of salt.

In two bowls, separate the raw egg yolks from the white. You can use the shell technique of going back and forth between two halves of a cracked shell or use your hands. For the latter, clean your hand well then crack open the egg with one hand into the other over a bowl and let the egg white slip through your fingers. Add the separated egg yolk on the layer of salt. Give space between each egg yolk so they are not touching.

When all egg yolks are on salt bed, cover them with more salt until you can't see much yellow.

Now wait for 3 days and unbury the egg yolks. You now have salt-cured egg yolks. Store them in a fridge indefinitely—without water and with high salt content, they are shelf-stable. Grate them on foods as a savory, salty spice.

Salt

Salt has a greater impact on flavor than any other ingredient. It enhances many basic tastes, suppresses bitterness, and increases aromatics (see Fig. 9.6, Taste Interactions). In addition, it has a basic taste of "salty," which the body innately likes at lower concentrations. When a food "tastes" flat, the culprit is often under seasoning with salt. Salt is a mineral composed of two elements, sodium (Na) and chloride (Cl). Diffusion of salt is slow when not quickened by heat or water. For this reason, you want to salt some big, dense foods (like meats) days in advance to allow time for it to penetrate to the center. Remember, when you season from the inside, each bite becomes enjoyable. For moist foods, such as fish, a few minutes before cooking is enough to season the center as the water content promotes diffusion. Cooking time is also an important factor. For quick cooking times, as seen with boiling pasta or blanching vegetables, the water needs to be salted more aggressively than for longer cooking times, as seen with boiling grains.

All cooking salt is salt (NaCl), whether mined from ancient lakes or dried from modern oceans, but they have different sizes and impurities. The size of the salt matters a lot in cooking, as the same volume of larger salt (e.g., fleur de sel), compared to smaller (e.g., table salt), will salt the food less but add more texture when finishing a dish. There is also a big difference in price, as manufacturing of different sizes has different costs. Fleur de sel is large flake salt from careful monitoring of specific environmental conditions and precise harvesting methods during the salt production season in western France, which typically lasts for a few months each year. Remember: small or large flaked salt will dissolve in water similarly; therefore, salting a sauce or pasta water with larger, more expensive salt is a waste. Therefore, it's recommended to use table salt or kosher salt for cooking (small to medium size) and larger flakes for finishing a dish. Larger salts can enhance the flavor of a finished dish, adding a "crunchy" texture or "salty" pop to a dish.

Salt's role in cooking extends beyond flavor; it profoundly impacts the texture and structure of foods by interacting with proteins. Proteins, the building blocks of meat, vegetables, and doughs, are chains of amino acids that respond to salt in different ways, depending on the food. In many foods, salt softens proteins, a process that is particularly beneficial for cooking both meats and vegetables. When you brine meat, for example, salt penetrates and interacts with muscle fibers, breaking down their structure and making the meat tender. Simultaneously, salt alters the protein structure in a way that increases its water-holding capacity. This ability to trap water inside meat explains why brined chicken remains juicy even after roasting. As salt diffuses into the meat, it draws in water through osmosis, effectively seasoning the meat from the inside while making it moister. The same principle applies to vegetables, which contain structural proteins in their cell walls. During blanching or quick cooking, heavily salting the cooking water has two critical effects. First, it helps season the vegetables from within, as their high-water content promotes rapid diffusion of salt into their tissues. This ensures that vegetables are seasoned from within. Second, salt disrupts the proteins in the cell walls of vegetables, softening their structure. This protein disruption accelerates the cooking process, allowing vegetables to become tender more quickly while maintaining a bright, appealing color. This is why chefs often emphasize heavily salting the water for blanching or boiling vegetables—it improves flavor, texture, and color!

Interestingly, salt takes on the opposite role in doughs and batters. When making bread, salt strengthens the gluten network, the web of proteins responsible for a dough's elasticity and structure. Gluten proteins, gliadin and glutenin, form a network when hydrated and kneaded, and salt encourages tighter bonds between these proteins. This reinforcement is crucial for trapping the gases produced by yeast fermentation, allowing bread to rise with a strong yet elastic structure. Without salt, dough can feel slack and fail to hold its shape, resulting in denser bread.

For some cooking applications, like baking, a scale can weigh the right amount of salt needed. However, to measure salt in most cooking methods, your senses are the best approximation. "Practice makes perfect" is a rightful saying for this situation. Tasting while cooking to gauge the right amount of seasoning in a food is a skill that takes practice. Find a salt type you like, then begin to pay attention to what works using your senses of taste, touch, and vision. Don't use a measuring spoon to add salt—get a feel for the salt amount that seasons. This is especially important as most recipes are not seasoned for those with limited or no smell. For instance, when salting a piece of meat, notice the weight of the salt in your hand and how it feels sprinkled with your palm or through your fingers. After salting, notice the coverage of white flakes to dark meat. Allow the salt to penetrate the center, cook, and taste the meat. If the taste is to your liking, commit these sensations to memory. If not, make a mental note to increase or decrease next time. Follow these steps for all cooking, fast or slow. Seasoning a stock? Taste the stock, handle some weight of salt, and sprinkle it in, wait for it to dissolve in the liquid then taste and salt again.

Monosodium Glutamate (MSG)

Monosodium glutamate (MSG) is a type of salt, formed as the reactionary product between an acid (glutamic acid) and a base (sodium). Like table salt (sodium chloride), MSG carries a charge and diffuses easily into food via water. However, it is chemically distinct from table salt and primarily used to enhance savory flavors rather than impart saltiness.

You may have heard negative stories about MSG, such as its association with "Chinese Restaurant Syndrome," a term originating in the late twentieth century to describe alleged symptoms like headaches or stomach aches after consuming Chinese food. However, scientific studies have not substantiated these claims, and many researchers now consider the term misleading and rooted in racial bias rather than evidence-based science. MSG is safe to consume in normal dietary amounts and is recognized by the U.S. Food and Drug Administration (FDA) as "generally regarded as safe" (GRAS).

MSG, like salt, is a compound naturally present in the body and in various foods. It is a powerful flavor enhancer, known for amplifying umami—the savory "fifth taste" alongside sweet, salty, sour, and bitter. Unlike salt, MSG is used in much smaller quantities due to its potency; a recommended starting point is around 0.1% of food weight, compared to 1% for salt. MSG may be purchased as a seasoning or derived naturally from ingredients like tomatoes, cheese, and seaweed. Alongside glutamates (like MSG), compounds such as inosinate and guanylate also enhance umami flavors and are found in certain foods (see Table 9.1).

Table 9.1 Foods naturally high in glutamates, inosinate, and guanylates

Glutamates	Inosinate	Guanylate
Seaweed (Kombu)	Dried Bonito flakes	Dried Shiitake mushroom
Soy sauce	Tuna	Enoki mushroom (heated)
Parmesan cheese	Pork	Shimeji mushroom
Dried shiitake mushroom	Chicken	Common mushroom
Tomatoes	Sardines	Nori (seaweed)
Anchovies	Cod	Green peas

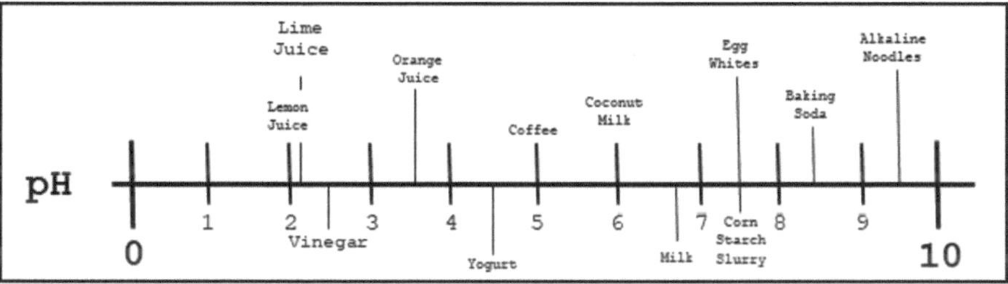

Fig. 9.5 Common agents of acid and bases used in cooking

Exercise

To experience MSG's flavor-enhancing power, try a simple experiment: Take three slices of ripe tomato at room temperature. Leave one unsalted, season the second with a pinch of table salt, and sprinkle the third with a much smaller pinch of MSG. Taste each slice. You'll notice the natural savoriness of the tomato in the first slice, enhanced by salt in the second, and amplified even further by the MSG in the third. Both salt and MSG elevate umami, but MSG delivers a distinct boost to the savory profile of foods.

Acidity in Cooking

Acids and bases are defined by their ability to shed or receive hydrogen ions (H+), respectively. In addition, bases may shed COOH. Strong acids/bases involve the entire chemical compound in this reaction, while weak acids/bases are only partially involved. The metric to measure acidity is its pH (or power to the H+) where an acid has a pH of 0 to 7 and a base between 7 and 14. Water has a neutral pH of 7. pH is a logarithmic scale of concentration of H+, a measure of acidity. Acids are more common in cooking than bases. Many foods contain several kinds of acids. For instance, egg whites are acidic, while their yolk is a base. Acids have a myriad of sources, in various forms that range in pH (Fig. 9.5).

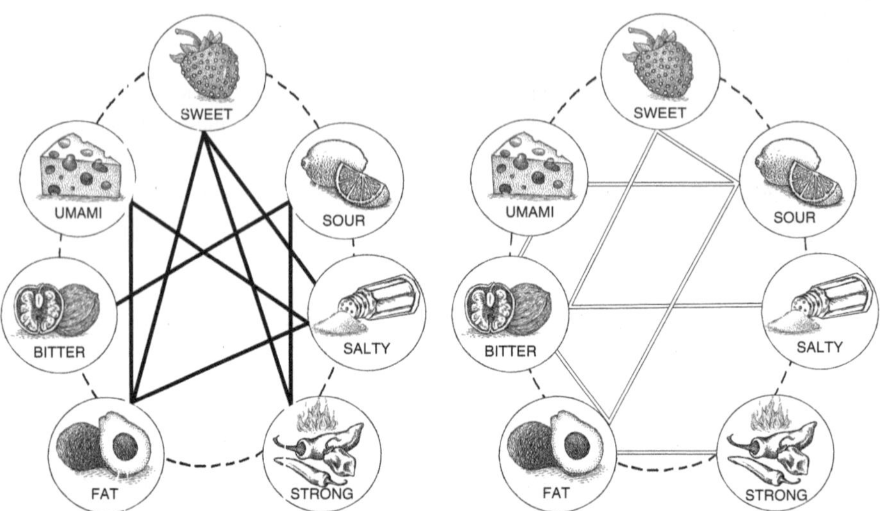

Fig. 9.6 Basic taste interactions for balancing tastes. The black lines (left) show which tastes enhance each other, while the white line (right) show which basic tastes balance (dampen) each other. Use these interactions when seasoning with basic tastes to balance a dish

Adding acidity in cooking is about adding balance to a dish. Some dishes have very forward acid notes like a lemon meringue, while others are subtle like a shot of sherry in a rich stew. The incorporation of acids can add lightness to a heavy dish and balance out sweetness (but also other tastes; see Fig. 9.6, Taste Interactions). Or splashing in one or more types of acid can contribute additional layers of complexity. Acids can change the shapes of proteins in food in a similar way as heat. While heat uses energy to break bonds, acids create a charge imbalance in the protein, leading to it unfolding and reforming. This chemical cooking is used to keep the whites of eggs together when poaching by adding vinegar to the simmering water or fully cooking fish in dishes like ceviche. Additionally, a simple squeeze of lemon (or another acid) can help prevent browning of sliced fruits and vegetables by neutralizing their surface enzymes. Basic solutions can also "cook" the food. The thousand-year egg, a Chinese delicacy, uses strong base lye to cook eggs. Fermentation and pickling are also methods to develop acidity in foods as well as modify their texture (Fig. 9.6).

In laboratories, pH is measured by devices that detect the H+ concentration. Papers that change color to indicate pH are similar but less accurate. If you're using acid to preserve or pickle foods, it's a good idea to use one of these methods to be safe. Here, a food with a pH under 4.6 is considered acidic, and pasteurizing it does not need temperatures above boiling water. For general cooking, the sensation on your tongue is the best instrument. Our body is extremely sensitive to acidity. Our blood has a pH of about 7.2 and minute variation can create havoc to the many proteins in it. Anything that tastes sour is acidic, falling under 7 on the pH.

Exercise
To dial in a range of acidity, use the salivating rate as a metric. Along with chewing, acid is one of the most effective ways of stimulating saliva. Apply some acids from Fig. 9.5 to your tongue, wait a few seconds, then lean your head forward. The lower the pH, the more saliva should be collected. With practice, you will be able to get a reasonable approximation of acidity. Apply this knowledge with continual tasting while adding acids to dishes, similar to training your palate to salting. In time, you'll get a sense of how each acid can be used to balance and complement different dishes.

Experiment in the Kitchen

We've provided you some basics on transforming your raw ingredients to cooked food. There is a lot more to learn, so use the Further Reading (at the end of the chapter) to guide you to several great books on the subject. Food science can help you deconstruct a recipe by understanding what the ingredients are made of and why certain methods of cooking are used. Remember: Recipes are instructions for cooking. Knowing why the original authors wrote these instructions will allow you to more easily rewrite them and add or substitute ingredients and modify cooking methods. In the process, you will create new flavor experiences. This is especially important, as almost no cookbooks are designed for those who cannot smell.

Cooking, like science, is a way of thinking. It's about being curious, making observations, and trying to understand those observations and applying this learned knowledge to new experiments in the kitchen. Learn to be a scientist in the kitchen; you'll not only make better food and eat better, but you'll continually be entertained by this engaging activity.

Individual Differences

What you can taste and sense when you suffer from loss of smell is very individual. Some people have a reduction in their sense of smell and do not feel they can register smells in their surroundings or the aroma of food, but despite this, they are still within the normal range when their sense of smell is tested with an established smell test. Others have virtually no measurable sense of smell but only feel a slight reduction in the flavor experience. Those who have combined loss of smell and taste, such as the late effects of COVID-19 or radiation/chemotherapy, often experience the additional complexity of which basic tastes and smells come back first and at what pace. In addition, saliva production may be reduced, which creates additional challenges and a need for increased focus on textures and the use of sour and umami, which stimulate saliva production. For these reasons, it is challenging to provide general guidelines on food preparation or identify specific recipes that offer the greatest enjoyment. However, there are compensation strategies that might bring back some enjoyment to the eating experience.

Compensation Strategies

Patients with chemosensory loss often adopt different strategies for coping with their altered eating experience, typically falling into two categories: those who consume less due to reduced interest in food and those who consume more to compensate for the diminished sensory signals. This behavior is rooted in the understanding that the sensory attributes of food are key drivers of food enjoyment. However, weight among individuals with smell loss often fluctuates or remains stable, suggesting that cognitive factors also play a significant role in eating behaviors.

To address these challenges, two distinct compensation strategies emerge:

1. *Bringing all senses to the table*—This involves evoking and enhancing the remaining sensorial aspects of food to create a richer eating experience. The most important advice is to try different foods—the rule of thumb is the more you feel in your mouth, the better. In addition, you should stimulate as many senses as possible.
2. *Shifting focus away from flavor*—This strategy shifts attention from flavor to other enjoyable aspects of eating, such as the social connections during meals or the creative process of cooking.

By combining these approaches, individuals can navigate the challenges of chemosensory loss while rediscovering the pleasures of food in new ways.

Bring All Senses to the Table

Research on loss of smell shows that patients most frequently report reduced enjoyment of food as a major problem. Many patients describe how they have completely given up on the pleasure of a meal. The patients' problem has thus been clearly defined for many years, but research has largely focused on improving the sense of smell and especially the results in smell tests. The problem with this approach is that not everyone can achieve sufficient improvement in the sense of smell to bring back the enjoyment of food.

The old saying "if one sense is lost, the others are sharpened" has unfortunately been shown to be invalid in several research studies. In fact, it is more often seen that the other senses (taste and mouthfeel) also take a small step in the wrong direction after a loss of smell. Part of the explanation for this is that the combined impressions of the senses are greater than the individual sensory impressions individually.

People with loss of smell often have increased attention to both taste and mouthfeel. However, other sensory impressions from the food, including appearance, crunchy sounds, and tactility/touch, rarely received more attention, and a large proportion of participants with loss of smell even had reduced attention to the sense of taste after their loss of smell. Many participants with reduced taste awareness described giving up on enjoying food in general after their loss of smell. Although mouthfeel, taste, and smell are important to our experience of food, our other senses can contribute more to the meal than we usually think about. Our brains are designed to explore new potential food sources using other senses—just think of the first person who had the idea to eat an oyster or the person who came up with the idea for surströmming. You can gain new taste impressions and food experiences by using your eyes, ears, and hands in cooking and during the meal.

The Taste Kit

An effective way to regulate our behavior is to make the actions we want easy and accessible. So, establishing taste kits that are visible and within sight and reach will make you more aware of using the tools that help you in your daily cooking. Make sure to have tasters in the kitchen that contain the different basic tastes as well as a collection of foods that are crunchy and have other consistencies— preferably added to some of the foods you can taste/sense best. These can be, for example, some types of spices, different kinds of strong elements, and pickled crispy foods such as cornichons, pickled cucumbers, pickled onions, pickles, and the like. Remember, the five basic tastes cover a variety of foods (Fig. 9.7).

Here are some good examples of the contents of a Taste Kit that contains all five basic tastants and ingredients for spiciness (trigeminal stimulation):

Sour—Lemon, vinegar, balsamic
Salt—Flaked salt
Sweet—Honey, sugar/icing sugar
Bitter—Dijon mustard (also strong)
Umami—Tamari, good soy spicy
(trigeminal)—Pepper, Sambal olek/harissa, Tabasco

Optimizing the Dining Experience

The following concrete examples show effects that normally work for people without loss of smell and that can be used to positively support the dining experience. To what degree this is also the case for meals where patients suffer from loss of smell is not known and has not yet been studied. But it is indisputable that good memories and a nice atmosphere in the room and the dining situation have a positive effect, so give it a try.

Fig. 9.7 Basic tastants. Put together a kit with all the basic tastes by the stove, so that it is easily accessible and the seasoning is remembered every time you cook

Food with Many Colors

In the brain, colorful food equals nutrients. That is why we are attracted to that food and eat more of it. So think of all the colors of the rainbow in your cooking. For example, elevate your salad with different shades of green, red cabbage, yellow peppers, carrots, and colorful tomatoes.

Heavy Cutlery and Rough, Thick Napkins

Food we eat with heavy cutlery tastes better because a subconscious feeling of quality is transferred from the cutlery to the food. So, give the food more weight and set the table with the heaviest cutlery. A thick and rough napkin signals quality in the same way as heavy tableware.

Music

Our mood and state of mind have a huge impact on how good we think the food tastes. Therefore, listening to music that you like can be positive for your dining experience. Music is very mood-setting, so listen to something that brings back good memories before and during dinner. The music does not have to be loud to affect the mood—often it is low-pitched music that is most pleasant. Additionally, the concept of *sonic seasoning* suggests that certain types of music or sounds can enhance specific flavors. For example, high-pitched, tinkling sounds might accentuate sweetness, while deeper, more resonant tones may bring out bitterness. Carefully selecting music that complements the flavors of your meal can further heighten the sensory experience.

Dark Plates

Food served on dark and black plates creates a greater color contrast and looks more inviting—especially if the food itself on the plate also contains many colors. The visual impression of food is very crucial for how the taste is perceived and how strong it tastes, so by intensifying the contrasts, the taste will also be perceived as stronger.

Focus On the Food

Studies have shown that if we do not focus on the food and our senses are instead used to watch TV, for example, then the feeling of fullness lasts shorter. You eat with all your senses, and it is therefore important to focus on presence for optimal dining pleasure. A beautiful presentation of the food can help to emphasize the focus on the food and can in itself also enhance the dining experience.

Lighting

We eat slower and are more relaxed when the lighting is slightly dimmed and not too bright. Think about the lighting in the room where you eat. Feel free to light candles and make sure there is a nice atmosphere in the room. But don't overdo it; the colors and visual temptation of the food is important too. For instance, we tend to generalize intensity ratings, so low lighting may lessen the intensity of other sensory aspects of food, such as taste.

Recall Memories Through Food

Most people have a wealth of dishes that they associate with places or people. It may be something you had in your childhood, at your grandparents' or on trips around the world. It can be rewarding for a dining experience to recall memories and good memories.

Avoid Annoying Background Noise

It is a good idea to turn off sounds from the hood, dishwasher, and TV during the meal. This way, any noise not adding to the sensory enjoyment of food is reduced as much as possible during the meal. In fact, background noises often lead to less ability to discriminate differences among foods.

Shifting Focus Away from Flavor

We humans are inherently social creatures, and meals often taste better when shared with others. Research has shown that shared experiences amplify emotions and perceptions, making pleasant moments more enjoyable (but also unpleasant ones more intense). This phenomenon, called the amplification effect, highlights the power of social connection during mealtimes. For individuals coping with smell loss, dining with others can add richness and depth to the experience, helping to compensate for the diminished sensory input.

Social contexts, especially during main meals like dinner, become integral when flavor takes a back seat. While breakfast may feel routine and solitary, sitting down with loved ones at dinner can foster feelings of connection and shared joy. Even without explicit communication, simply knowing that someone else is enjoying the same meal amplifies the experience. This shared psychological focus draws more attention to the meal, enriching its overall enjoyment.

In addition to eating together, the act of cooking offers another valuable compensation strategy. Cooking shifts the focus from the outcome (eating) to the process itself—a creative and rewarding activity. When cooking is shared, such as preparing a dish with a friend, partner, or family member, it takes on new meaning. It becomes an opportunity to bond, collaborate, and create shared memories, making the process as fulfilling as the meal itself. Social and cooking experiences remind us that meals are not just about taste—they are opportunities for connection, exploration, and joy.

Effect of Interventions Toward Regaining Pleasure of Food After a Smell Loss

While the effects of smell loss on the perception and pleasure of food have been examined in numerous studies with a wide range of outcomes, only a few studies have investigated interventions. The effects of capsaicin, the molecule responsible for the spiciness of chili, have recently been explored in patients with smell loss. Researchers found that for one of their food samples, adding a moderate amount of spiciness had a small positive effect on overall liking while generally increasing perceived flavor intensity. In another study, the effects of capsaicin were compared with a control group with a normal sense of smell, where neither the effects of capsaicin, cooling menthol, changes in viscosity nor the addition of umami had any superior effects in patients with smell loss compared with the control group. So far, there is no compelling evidence that enhancing a single sensory element in a meal can produce large effects in the compensation of smell loss on flavor perception.

In a Danish study, patients with smell loss participated in a series of cooking school classes to investigate if a more comprehensive intervention could change the pleasure of food and cooking.

During the five-week cooking school, participants were introduced to a broad knowledge of how the senses work and interact during eating with numerous tasting samples for all flavor senses to facilitate the strategy of learning through the senses. With guidance from chefs, participants cooked flavor-enhanced dishes before presenting them during the social dining at the end of each session. Through the feedback from participants, the contents and recipes of the cooking school were continuously adjusted. This included the overall 6T framework, where tools such as the Taste Kit were developed to enhance the balancing of the basic tastants, and a Texture Bank was developed to ease the inclusion of different premade texture-enhancing ingredients. The knowledge gained from these cooking schools has been used for many of the recipes in this book.

The 6T rule to create palatable dishes for individuals with olfactory loss

Sensation	Description
Taste	Use all five basic tastants to ensure proper balance and flavor of the meal before serving and during eating.
Trigeminal Strength	A bit of strength on the tongue adds to the overall enjoyment of the meal, but too much can destroy the meal for both individuals with a normal and affected sense of smell.
Temperature	Variations in temperature in the dish adds to the multisensory experience of the meal.
Texture	A successful meal must include several different textures, such as creamy, chewy, crisp, crunchy, moist, tender, sticky, soft, lumpy, rough, liquid, semi-liquid, gritty, crumbly, and smooth.
Touch	The sensory experience of a meal is initiated before the food enters the mouth. By making it visually appealing and including elements eaten without cutlery, the fingers and vision start the multisensory cascade of input to the brain before other senses are activated. Giving a meal the final touch before serving it, can set up expectations and add to the sensory complexity of a meal.
Temptation	Vision and the smell of cooking and food induce appetite, saliva production, and attention on food. If the sense of smell is not fully lost, an increase of smells can be beneficial. An esthetically arranged plate and table can boost the pleasure from the meal.

During cooking in groups, the participants with smell loss could use the new knowledge of all senses contributing to flavor and share experiences and issues regarding their smell loss in a relaxed atmosphere. All dishes were discussed and evaluated during the social dining to improve the recipes and to promote the discussion and sensory evaluation of the meal. During the last session, participants brought relatives to the cooking school, who, after a short introduction to the principles and senses, cooked and dined with the other participants to promote the concept and with a hope to expand the continuous cooking experience to kitchens at home.

There were several positive effects of this combined multisensory and social intervention. Participants reported an improvement in their quality of life, became more involved in daily cooking, improved confidence in their abilities to choose ingredients and handle unexpected results, found cooking to be a meaningful activity, and preferred to cook their meals instead of having others cook for them. Compared with before the cooking school, participants showed both improved tested and self-rated smell function at the three-month follow-up after the cooking school, where the positive effects on involvement in daily cooking and positive attitude toward cooking remained high.

Additional Reading

Journalistic Articles

Steingarten, Jeffrey. "Why Doesn't Everyone in China Have a Headache?" *Vogue*, 1999. Reprinted in *It Must've Been Something I Ate: The Return of the Man Who Ate Everything*, Vintage, 2002.

Reference Cookbooks

Katz, S. E. (2012). *The Art of Fermentation: An In-Depth Exploration of Essential Concepts and Processes from Around the World*. Chelsea Green Publishing.

Fjaeldstad, A., Bredahl, R. & Bøjlund, C. Cooking With a Smell Loss. Apple Books, 2023, Google Play 2025.

Myhrvold, N., Young, C., & Bilet, M. (2011). *Modernist Cuisine: The Art and Science of Cooking*. The Cooking Lab.

McGee, H. (2004). *On Food and Cooking: The Science and Lore of the Kitchen*. Scribner.

Textbooks and Academic Resources

Chang, R., & Goldsby, K. (2016). Chemistry (12th ed.). McGraw-Hill Education.

Civille, G. V., & Carr, B. T. (2015). Sensory Evaluation Techniques (5th ed.). CRC Press.

Damodaran, S., & Parkin, K. L. (Eds.). (2017). Fennema's Food Chemistry (5th ed.). CRC Press.

Braude, L. & Stevenson, R. J. Watching television while eating increases energy intake. Examining the mechanisms in female participants. Appetite 76, 9–16 (2014).

Ferris, A. M., & Duffy, V. B. (1989). Effect of olfactory deficits on nutritional status: does age predict persons at risk?. Annals of the New York Academy of Sciences.

Fjaeldstad, A. W., & Smith, B. (2022). The Effects of Olfactory Loss and Parosmia on Food and Cooking Habits, Sensory Awareness, and Quality of Life—A Possible Avenue for Regaining Enjoyment of Food. *Foods, 11*(12), 1686. https://doi.org/10.3390/foods11121686

Fjaeldstad, A. W. Using Cooking Schools to Improve the Pleasure of Food and Cooking in Patients Experiencing Smell Loss. Foods 13, 1821 (2024).

Fjaeldstad, A. W. Culinary Cure? Improved Subjective and Measured Sense of Smell after a Cooking Rehabilitation Program for Patients with Smell Loss. International Journal of Gastronomy and Food Science, Volume 39, 101121 (2025).

Higgs, S., & Ruddock, H. (2020). Social influences on eating. Handbook of eating and drinking: Interdisciplinary perspectives, 277–291.

Hunter, S. R., & Dalton, P. H. (2022). The need for sensory nutrition research in individuals with smell loss. *Clinical Nutrition Open Science*, *46*, 35–41. https://doi.org/10.1016/j.nutos.2022.11.002

Landis, B. N., Scheibe, M., Weber, C., Berger, R., Brämerson, A., Bende, M., Nordin, S., & Hummel, T. (2010). Chemosensory interaction: Acquired olfactory impairment is associated with decreased taste function. *Journal of Neurology*, *257*(8), 1303–1308. https://doi.org/10.1007/s00415-010-5513-8

Michel, C., Velasco, C. & Spence, C. Cutlery matters: heavy cutlery enhances diners' enjoyment of the food served in a realistic dining environment. Flavour 4, 26 (2015).

Pellegrino, R., & Fjældstad, A. (2024). The Effect of Olfactory Disorder (and Other Chemosensory Disorders) on Perception, Acceptance, and Consumption of Food. In Smell, Taste, Eat: The Role of the Chemical Senses in Eating Behaviour (pp. 119–137). Cham: Springer International Publishing.

Pellegrino, R., Hummel, T., Emrich, R., Chandra, R., Turner, J., Trone, T., et al. (2020). Cultural determinants of food attitudes in anosmic patients. Appetite, 147, 104563.

Pellegrino, R., Luckett, C. R., Shinn, S. E., Mayfield, S., Gude, K., Rhea, A., & Seo, H. S. (2015). Effects of background sound on consumers' sensory discriminatory ability among foods. Food Quality and Preference, 43, 71–78.

Philpott, C. M., & Boak, D. (2014). The Impact of Olfactory Disorders in the United Kingdom. *Chemical Senses*, *39*(8), 711–718. https://doi.org/10.1093/chemse/bju043

Reinbach H. C., Riantiningtyas R. R., Bredie W. L. P., Fjaeldstad A. W. Effect of trigeminal stimulation, texture and taste enhancement on food pleasure in adults with olfactory dysfunction (Article in prep).

Practical Cookbooks

Arnold, D. (2014). Liquid Intelligence: The Art and Science of the Perfect Cocktail. W. W. Norton & Company.

Brenner, M., Sörensen, P., & Weitz, D. (2020). Science and Cooking: Physics Meets Food, From Homemade to Haute Cuisine. W. W. Norton & Company.

Corriher, S. O. (1997). CookWise: The Hows and Whys of Successful Cooking. William Morrow.

Goldwyn, M. (2016). Meathead: The Science of Great Barbecue and Grilling. Rux Martin/ Houghton Mifflin Harcourt.

Kamozawa, A., & Talbot, H. A. (2010). Ideas in Food: Great Recipes and Why They Work. Clarkson Potter.

Nosrat, S. (2017). Salt, Fat, Acid, Heat: Mastering the Elements of Good Cooking. Simon & Schuster.

Potter, J. (2010). Cooking for Geeks: Real Science, Great Hacks, and Good Food. O'Reilly Media.

Ruhlman, M. (2009). Ratio: The Simple Codes Behind the Craft of Everyday Cooking. Scribner.

Vega, C., Ubbink, J., & van der Linden, E. (Eds.). (2012). The Kitchen as Laboratory: Reflections on the Science of Food and Cooking. Columbia University Press.

Part IV
Recipes

Chapter 10
General Introduction to the Recipes

Contents

Normally, cookbooks and recipes are made with an attempt to create a nice sensory experience, often with a focus on specific aspects of cooking such as classical culinary cooking, vegetarian diets, weight loss, or optimizing techniques and usage of specific equipment such as a barbeque or an air fryer. These recipes are made under the assumption that all senses are intact and that the reader is able to perceive the ingredients and dish in the same way as the chef or author. After experiencing smell disorders and knowing the commonality of smell (and other sensory) impairments, this is clearly not the case.

In this final section of the book, the sensory knowledge and compensatory strategies are merged with methods to improve overall sensory enjoyment for a list of recipes made specifically for individuals suffering from smell disturbances. We highlight how tools, methods, and ingredients emphasize the basic tastes, tactility, temperature differences, trigeminal spiciness, texture, and temptation to eat the dish. This is not only to improve the sensory output perceived for this single dish but can be used to improve general understanding of a new way to think of cooking. Think of it as sensory training using building blocks of recipes and sensory perception to open a new way of cooking and enjoying food.

Many of the principles and recipes were created during a cooking school project for patients with smell loss, which have been thoroughly tested by several classes of patients with smell loss. Our previous experience in both research and clinical work with smell loss combined with feedback provided by these participants on the cooking schools has given us a unique insight on how to provide a sensory boost to cooking and improve culinary skills, culinary confidence, and increased pleasure of both

A. W. Fjaeldstad et al., *Rediscovering Flavor*,
https://doi.org/10.1007/978-3-032-08056-1_10

Fig. 10.1 The 6T's. Several other senses can add to the pleasure of food. These become even more important if the sense of smell is reduced or absent. Use the 6T's to boost the sensory experience of your food

cooking and eating. By applying simple tools such as the taste kit to balance the basic tastes in a meal and using the 6T's to enhance sensory stimulation (Fig. 10.1), you can reboot the way you cook, where a smell impairment does not limit your abilities to create a flavorful and delicious meal.

How to Approach the Recipes

These recipes are designed to help you rediscover the joy of cooking. Be sure to read each recipe thoroughly before you begin. To fully understand a recipe, focus on balancing flavor through the 6Ts: *Taste, Texture, Temperature, Tactility, Temptation,* and *Trigeminal Sensation*. As introduced in earlier chapters, these principles provide a framework for enhancing the sensory aspects of each dish. By consciously engaging with these elements, you can transform cooking into a vibrant, sensory-rich experience that brings pleasure back to the kitchen and dining table.

Reading a Recipe

Understanding the recipe format is essential for maximizing your culinary experience and tailoring dishes to your liking. Each recipe begins with the name of the dish, followed by preparation and cooking times and the number of servings—giving you a clear sense of the time and effort involved. A brief introduction highlights the recipe's appeal through its textures, temperatures, and tastes, emphasizing sensory experiences beyond smell.

The *Sensory Profile* (used for main dishes) offers insights into the dish's characteristics based on the Six Ts. Use these profiles, along with tasting the final dish, to assess your sensory experience. For example, if you enjoy spicy foods but the recipe has a low trigeminal score (1 out of 3 chili peppers), you might consider adding more spices. However, it's recommended to try the original recipe first to establish a baseline for the flavors before making adjustments.

Sensory Profile (based on 6 Ts)
TASTE ■■■■□ *Sweet* ■■□□□ *Salty* ■■□□□ *Bitter* ■□□□□ *Sour* ■□□□□ *Umami*
TEXTURE ■■■■□ *Creamy* ■■■□□ *Crunchy* ■□□□□ *Chewy*
TEMP ❄ *Cold* | 🌡 *Room Temp* | 🔥 *Hot dish* *TRIGEMINAL* 🌶🌶□
TACTILE ✋ *Hand-held* | 🍴 *Utensils* *TEMPTATION* ■■■□□

Ingredients are listed in the order they are used, with precise measurements and any necessary preparation details provided. Methods describe the cooking in a clear, step-by-step format, often grouped logically to make the process easy to follow. Suggestions for assembly and presentation guide you in plating the dish to enhance its visual appeal and sensory impact, emphasizing the temptation aspect of the 6Ts.

Sensory Notes delve deeper into how ingredients and cooking methods shape the sensory experience. They explain how specific ingredients influence taste, such as how a squeeze of lemon juice can brighten the flavors with acidity or how slow cooking results in a tender texture.

Alternatives offer ideas for customizing the dish further, while the *Cook's Notes* section encourages you to document your own adjustments, observations, and sensory experiences. Embracing experimentation and noting your experiences can turn cooking from a routine task into a creative and engaging activity.

Tailoring a Recipe to Your Liking

Personalizing recipes can deepen your connection to the food and increase your enjoyment by aligning the dish with your sensory preferences. Depending on your sensory loss, you may experience these recipes differently. As you gain more experience, you'll naturally develop ideas for modifications based on your preferences and how various ingredients affect the dish.

Keeping sensory notes as you cook will help you understand what enhances your flavor experience. The 6Ts framework serves as a helpful guide as you explore. Additionally, we recommend the *Taste Kits* and *Crunch Kits* from a cooking school for patients with smell loss to help you adapt each recipe to fully stimulate your senses. The Taste Kits include ingredients that enhance the dish's basic flavors, while the Crunch Kits provide elements that add texture and variety.

Taste Kit

To ease the process of achieving a good balance between basic tastes and remembering to flavor the meal, having a taste kit next to the stove can be highly recommended. However, some elements should be stored in the refrigerator to maintain flavor and increase shelf life. Make your own preferred taste kit, where you add sweet (honey or sugar), salty (salt and flake salt), sour (apple vinegar), bitter and trigeminal (mustard, consider refrigerator), and umami (tamari or good soy sauce) (see Fig. 10.2).

Fig. 10.2 The taste kit

Crunch Kit

Adding a bit of crunch to a soup, salad, or any other meal adds to the sensory stimulation and eating experience. Crisp or crunchy elements can be brought or made in advance. We have added a list of ideas for crunchy supplements, which can be stored for weeks in an airtight container.

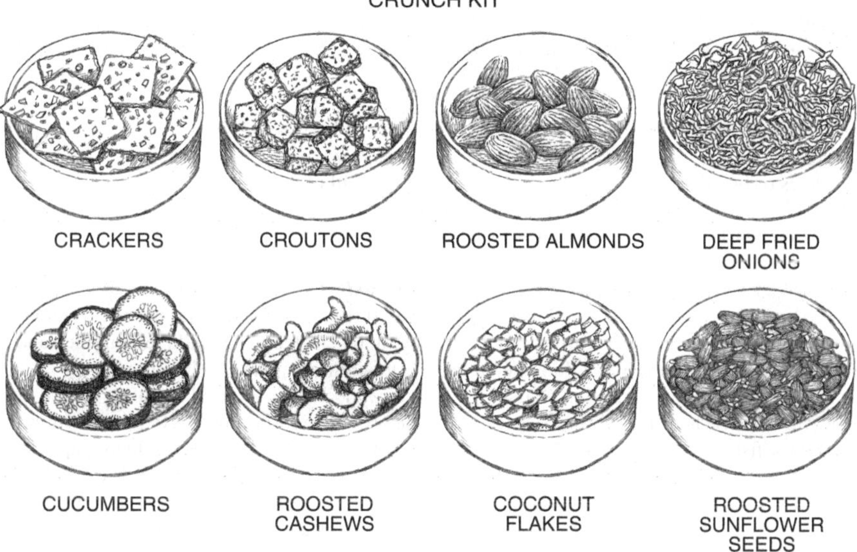

Rye Bread Croutons	**Whole Grain Croutons with Smoked Paprika**
Ingredients 2–3 slices of rye bread (or whole-grain bread) 1½ tsp. cane sugar ½ tsp. fennel seeds Coarse salt *Method* Slice the bread into small cubes or strips. Drizzle a little olive oil over and sprinkle with cane sugar, fennel seeds, and a pinch of salt. Bake in a preheated oven at 170 °C (340 °F) for about 10–12 min until crisp. Store in an airtight container and use as a topping for salads.	*Ingredients* 2–3 slices of stale whole-grain bread 1 tbsp. olive oil 1 tsp. smoked paprika Coarse salt *Method* Cut the bread into pieces. Drizzle a little olive oil over and sprinkle with smoked paprika and a bit of salt. Bake in a preheated oven at 170 °C (340 °F) for 10–12 min until crisp. Store in an airtight container and use as a topping for salads.

Toasted Coconut Flakes *Ingredients* 150 g coconut flakes 2 tbsp. Lemon juice 2 tbsp. powdered sugar *Method* Toss the coconut flakes with lemon juice and powdered sugar. Put them in a baking dish lined with parchment paper, and bake them in a preheated oven at 140 °C (280 °F) for about 35 min until they are light brown and crispy. Store in an airtight container and use for salads and desserts.	**Toasted Buckwheat Groats** *Ingredients* 1 bag of whole buckwheat *Method* Bake the buckwheat in a convection oven at 150 °C (300 °F) for about 45 min until light brown and crispy. Store in an airtight jar and use as a topping for spreads and salads.
Herb Breadcrumbs *Ingredients* 2–3 slices of bread 1 handful of mixed herbs such as basil, parsley, and tarragon (Rinsed and dried in a tea towel) 1 tbsp. olive oil *Method* Cut the bread into small pieces and place in a food processor along with the rinsed and dried herbs. Pulse whilst pouring in a thin stream of olive oil until you have somewhat coarse breadcrumbs. Use herb breadcrumbs as a topping for fish or meat.	**Curry Cashews** *Ingredients* 1 handful of cashews 1 tbsp. curry powder 1 tbsp. lemon juice Coarse salt *Method* Roast the cashews in a pan over medium heat with the curry powder. Add lemon juice and salt, stir around, and let it simmer. Once the nuts have dried out, they should now be coated in curry powder. Place everything on a baking tray lined with parchment paper and let it dry out for about 5 min at 130 °C (265 °F) in a preheated oven. Use curry cashews as a topping for Indian or Thai salads or eat them as a snack. Tip: You can use almonds, sunflower seeds, and pumpkin seeds with the same method.
Tamari Almonds *Ingredients* 1 handful of almonds 2–3 tbsp. tamari (or high-quality soy sauce) *Method* Roast the almonds in a pan over medium heat then add tamari, or soy sauce and let it simmer. When the liquid has reduced, place the almonds on a baking sheet lined with parchment paper and dry it for about 5 min at 130 °C (265 °F) in a preheated oven. In taste and consistency, tamari almonds resemble bacon: use them on salads, egg dishes, or as a snack. You can also make sunflower seeds and pumpkin seeds using the same method.	

Exercise: Layering Senses into a Dish

This is a good exercise to do at home, where the effect of good flavoring truly shines.

The principles of multisensory flavoring apply to all types of dishes, but here they are shown for mushroom tartare

Mushroom tartarewith its high umami content, can be a good base for a dish but also works well as a side dish.

1. *Finely chop the mushrooms and sauté them until golden brown.*
2. *Flavor all the mushrooms in three steps and leave a portion of the mushrooms at each step on a plate, so they can be tasted against each other afterward:*

	Flavoring	Sensory stimuli	Tools
Step 1	Salt	Salty	Traditional flavoring
	Pepper	Spicy stimuli	
Step 2	+		+ Balanced flavoring with basic tastes
	Honey	Sweet	
	Apple Vinegar	Sour	
	Chives	Bitter	
	Mustard	Bitter/spicy	
Step 3	+		+ Flavoring with texture (+ visual preparation of the plate)
	Olive oil	Fat/creamy	
	Dried cranberries	Chewy	
	Nuts	Crispy	
	Roasted onions	Crunchy	

Chapter 11
The Recipes

Contents

A. W. Fjaeldstad et al., *Rediscovering Flavor*,
https://doi.org/10.1007/978-3-032-08056-1_11

Dips/Sauces

The following dips can be prepared days in advance, making them a convenient and versatile addition to your kitchen repertoire. Enjoy them with tortilla chips, veggie sticks, or as a spread on pita bread or pizzette. They're also fantastic for layering into sandwiches, topping grain bowls, or even as a quick fix for elevating leftovers. I almost always keep dips, along with things to dip, like crackers or crudités, and spiced nuts on hand—not just for impromptu game nights but also as a standby snack to keep diners' content during those longer-than-expected cook times for the main meal. You'll find that several of these dips overlap beautifully with the sides and snacks in the next section, offering endless ways to mix, match, and savor.

Beetroot Hummus

With its intense pink hue and balanced flavors, this hummus is as much a treat for the eyes as it is for the palate. The earthy sweetness of beetroot, nutty tahini, and tangy lemon create a well-rounded taste, while optional spice adds depth.

Ingredients
2 boiled beetroots, peeled & diced
1 can of chickpeas, drained & rinsed
15 mL (3 tbsp) tahini
10 mL (2 tbsp) olive oil
1 clove of garlic, pressed
½ lemon, juiced

Salt & pepper, to taste

Cooking Instructions
Blend all ingredients until smooth & creamy. Adjust seasoning with salt, pepper, & optional cayenne.

Assembly/Presentation
Garnish with toasted sesame seeds & fresh parsley.

Serve chilled with pita bread, veggie sticks, or chips.

Taste: The beetroot sugars are balanced by the lemon juice, and the Tahini contributes a rich, umami profile.
Texture: Smooth & creamy.
Temperature: Best served chilled—refreshing contrast to warm pita or crunchy veggies.
Tactility: Soft & scoopable—perfect for dipping with chips, veggies, or bread.
Temptation: Vibrant pink color with green parsley & golden sesame seeds to enhance presentation.
Trigeminal: Optional cayenne or chili.

Alternations
Increase color contrast: Mix in fresh mint or dill for an aromatic lift.
Milder Version: Reduce garlic for a softer flavor.
Boost nuttiness: Swap tahini for almond or cashew butter.

Cook's Notes: ____
____

Avocado and Lime Cream

This dip combines the mild and creamy avocados with the fresh acidic lime, which, apart from a good balance, preserves the vivid green color. For individuals with smell distortions, the onion and garlic of a traditional guacamole can be a strong trigger of parosmia. This is rather unlikely for this recipe, while still keeping the upsides. But be sure to make enough; this dip has a habit of disappearing fast.

Ingredients
1 poblano pepper
1 ripe avocados, peeled
4 tomatillo, quartered
1 garlic clove, pressed
125 ml (1 cup) crema
2 limes, juice and zest
5 mL (1 tsp) salt

Cooking Instructions
Roast poblano pepper in a hot pan, rotating to blacken all sides. Cool, then cut lengthwise and remove stem and seeds.

Blend poblano, avocado, tomatillos, garlic, crema, lime juice plus their zest, and salt on high until smooth.

Assembly/Presentation
Serve chilled topped with chili flakes with tortilla chips, veggie sticks, or with tacos.

To serve warm, place in a casserole dish and top with parmesan (or cotija) cheese and ½ that amount of chili pepper, then bake for 20 min at 200 C (400 F).

Taste: Fresh and diverse acidity from lime and tomatillo.
Texture: Avocado, along with crema, create a smooth texture.
Temperature: Serve chilled with warm tortilla chips or heat up dip.
Tactility: Take a dip with tortilla chips or drizzle it on a taco.
Temptation: Vibrant green hue is eye-catching and fresh—contrast this with red pepper.
Trigeminal: Mild heat introduces subtle, warming tingle on the tongue.

Alternations
Make parosmia friendly: Omit garlic which is a common trigger.
Extra Freshness: Stir in chopped mint or cilantro for an herbal touch.
Different Spice: Replace poblano with other hot green peppers (jalapeno, serrano).

Cook's Notes: ____
____

Mango and Chili Salsa

This vibrant, tropical salsa balances juicy mango sweetness, zesty lime acidity, and fiery chili heat. It's a versatile dip or topping that enhances everything from tortilla chips to grilled meats and tacos.

Ingredients
1 ripe mango, peeled and diced
1 red chili, finely chopped
½ red onion, finely chopped
1 handful fresh coriander, roughly chopped
1 lime, juiced
Salt, to taste

Cooking Instructions
Combine all ingredients in a bowl and gently mix.

Season with salt and adjust lime juice to taste. Let the salsa sit in the fridge for at least 30 min to allow flavors to meld.

Assembly/Presentation
Serve chilled as a dip for tortilla chips or a topping for tacos, grilled dishes, or salads.

Taste: Ripe mango brings natural tropical sweetness balanced by the lime juice.
Texture: Chunky and soft.
Temperature: Serve chilled for a refreshing contrast to warm dishes.
Tactility: Perfect for scooping, dipping, or topping various dishes.
Temptation: A vibrant mix of golden mango, red chili, and green coriander creates eye-catching contrast.
Trigeminal: Chili delivers a warming heat that can be adjusted for intensity.

Alternations
Increase creaminess: Mash some of the mango for a smoother consistency.
Reduce onion potency: Soak the red onion in lime juice for a softer taste.
Increase savoriness: Add diced cherry tomatoes for a tangy contrast.

Cook's Notes: ____
____

Sun-Dried Tomato Pesto

This rich and savory pesto features intensely sweet and umami-packed sun-dried tomatoes, nutty pine nuts, and fragrant basil. It's a versatile condiment that works as a spread, dip, or pasta sauce, adding depth and vibrancy to any dish.

Ingredients

100 g (½ cup) sun-dried tomatoes (in oil), drained
50 g (¼ cup) pine nuts
1 garlic clove
1 handful fresh basil leaves
½ dl (¼ cup) olive oil
Salt & pepper, to taste

Cooking Instructions

Toast the pine nuts in a dry pan over medium heat until lightly golden and fragrant. Let cool slightly.

In a food processor or blender, combine the sun-dried tomatoes, garlic, toasted pine nuts, and basil. While blending, slowly add olive oil, creating a coarse texture.

Season to taste with salt, pepper, and optional cayenne pepper for extra heat.

Assembly/Presentation

Serve as a dip, spread, or sauce for bread, pasta, or grilled vegetables. Swirl into soups for an umami boost.

Taste: Sun-dried tomatoes provide an intense, caramelized sweetness with a deep umami note.
Texture: Coarse and rustic in feel.
Temperature: Served at room temperature or cold.
Tactility: Perfect for spreading, dipping, or stirring into dishes.
Temptation: Deep red color, garnish with fresh basil for a green contrast.
Trigeminal: Cayenne or fresh chili is optional.

Alternations

Increase umami: Stir in grated Parmesan for added depth.
Increase acid: Add lemon juice and its zest for a bright, tangy contrast.

Cook's Notes: ____
____

Tzatziki

This refreshing, creamy dip combines cool cucumber with rich Greek yogurt, zesty lemon, and garlic for a perfectly balanced flavor. Ideal for pita bread, grilled meats, and fresh vegetables, tzatziki is a versatile and timeless addition to any meal.

Ingredients
1 cucumber, grated and squeezed dry
400 mL (1⅔ cups) Greek yogurt
2 cloves garlic, finely minced
Juice from ½ lemon
15 mL (1 tbsp) high-quality olive oil
0.6 mL (⅛ tsp) cayenne
Salt and pepper, to taste

Cooking Instructions
Salt the cucumber with 5 g (1 tsp) of salt, and let it sit for 15 min to promote moisture release.

In a bowl, whisk together the yogurt, garlic, lemon juice, and olive oil until smooth. Squeeze out excess moisture using a clean tea towel or paper towel, then stir into yogurt mixture until evenly distributed throughout the mixture.

Season with salt, pepper, and cayenne pepper if using. Cover and refrigerate for at least 30 min before serving to allow the flavors to meld.

Assembly/Presentation
Garnish with fresh dill or a sprinkle of paprika for color, then drizzle high-quality olive oil. Serve chilled as a dip for pita bread, a sauce for grilled meats, or a dressing for Mediterranean bowls.

Taste: Slight acidity from the lemon juice and Greek yogurt.
Texture: Smooth and thick sauce with slight crispness from the cucumber.
Temperature: Coolness of the yogurt and cucumber, especially when served cold.
Tactility: Scoopable and spreadable, with a creamy yet slightly textured consistency.
Temptation: White, creamy dip with flecks of green cucumber. Garnish with paprika or dill for color contrast.
Trigeminal: Adjust garlic strength and cayenne pepper (optional) for a milder or bolder kick.

Alternations
Add cooling: Add fresh mint for a different herbal cooling.
Increase creamy: Use half yogurt, half sour cream for a richer texture.
Increase trigeminal: Stir in chili flakes or harissa for a spicy version.

Cook's Notes: ____
____

Classic Cheese Spread

In the southern parts of the United States, you will not attend many parties without some form of processed cheese dish, either molded into a ball or thinned into a sauce for dipping. The recipe meets in the middle of those textures with a spreadable cheese that is enjoyed on multigrain crackers. An easily adjustable recipe can incorporate most pickled items in your fridge.

Ingredients
250 ml (1 cup) of cream cheese, softened
175 ml (¾ cup) of sharp white cheddar cheese, shredded
15 ml (1 tbsp) of Worcestershire
15 ml (1 tbsp) olive brine
45 ml (¼ cup) red onion, minced
7.5 ml (1 ½ tsp) horseradish
1.24 ml (1/4 tsp) paprika
0.75 (⅛ tsp) cayenne
30 ml (2 tablespoons) of whole milk (3.5% fat)

Cooking Instructions
In a large bowl, combine all ingredients until evenly incorporated. Shape cheese mix into a ball or place in a ceramic ramekin. Refrigerate until set (at least 1 h).

Assembly/Presentation
Serve cold alongside crackers, toasted bread, or raw vegetables.

Taste: A balance of tangy, savory, and spicy.
Texture: Smooth and creamy with small bits of onion for contrast.
Temperature: Served cold for firmness but spreads easily at room temperature.
Tactility: Spreadable with a knife or scoopable for crackers and veggies.
Temptation: The rich, golden cheddar color contrasts beautifully with garnishes like herbs, nuts, or paprika.
Trigeminal: Pungency and heat are present from the cayenne and horseradish.

Alternations
Change consistency: More milk makes it looser and more spreadable; omitting milk makes it firmer, like a cheese ball.
Increase umami: Add a splash of fermented hot sauce or soy sauce.

Cook's Notes: ____
____

Spinach Dip

This spinach and ricotta purée features a striking green color with visible white flecks from the cheese. Blanching the spinach enhances its vibrant hue while preserving its texture and nutrients. Simmering the ricotta with garlic infuses it with subtle flavor before blending it into a smooth mixture. The result is a balanced combination of creamy and vegetal elements. Served cold, it offers a refreshing contrast, while heating enhances its richness.

Ingredients
1.5 kg (3 lbs) spinach
500 mL (2 cups) ricotta cheese
30 mL (1 T) prepared horseradish
2 garlic cloves, finely chopped
5 mL (1 tsp) white pepper
10 mL (2 tsp) salt

Cooking Instructions
Bring a large pot of heavily salted water to a boil (should taste salty). Wash spinach, then blanch in boiling water for 30 s or until it turns a brighter green. Remove with tongs and immediately transfer to an ice bath to stop the cooking process.

In a small pot, combine ricotta, heavy cream, horseradish, white pepper, and garlic. Heat on medium-high until it begins to simmer. Reduce heat to low and let it simmer for 15 min to develop flavor.

While simmering, use your hands to bundle up spinach and squeeze out as much moisture as possible, then roughly chop. In a blender, combine the chopped spinach and ricotta mixture. Pulse until a smooth purée forms.

For a cold dip, refrigerate until chilled.

For a warm dip, preheat the oven to 175 °C (350 °F) and heat for 15 min before serving.

Assembly/Presentation
Serve in a shallow dish or ramekin. Pair with toasted bread, crackers, or vegetable sticks. Garnish with a drizzle of high-quality olive oil.

Taste: A blend of mild, creamy ricotta and vegetable brightness from the spinach.
Texture: Smooth and creamy, yet slightly airy from the ricotta.
Temperature: Serve cold for a refreshing dip or warm for a richer mouthfeel.
Tactility: Scoopable with crackers, bread, or vegetables.
Temptation: The deep green color contrasts beautifully with white flecks of ricotta.
Trigeminal: Mild pungency from garlic.

Alternations
Increase richness: Stir in 125 mL (½ cup) of cream or butter before blending.
Boost acidity: Add a squeeze of lemon juice for brightness.

Cook's Notes: ____
____

Sides/Snacks

Vegetable Sticks and Chips

Texture and color play a starring role in how we perceive and savor every bite, especially when it comes to snacks. This recipe showcases the humble vegetable stick—a versatile canvas for exploring vibrant colors and satisfying textures, as well as a perfect vehicle for many of the dips introduced earlier. Fresh and crisp vegetables bring a rainbow of hues to your plate, and changing their shape transforms how we experience them. Whether you're adding a delicate crunch or creating a bold bite, the shapes illustrated below demonstrate how a simple cut can redefine an ingredient. You can also experiment with vegetable cuts in other parts of your meal, adding texture and dimension to main dishes.

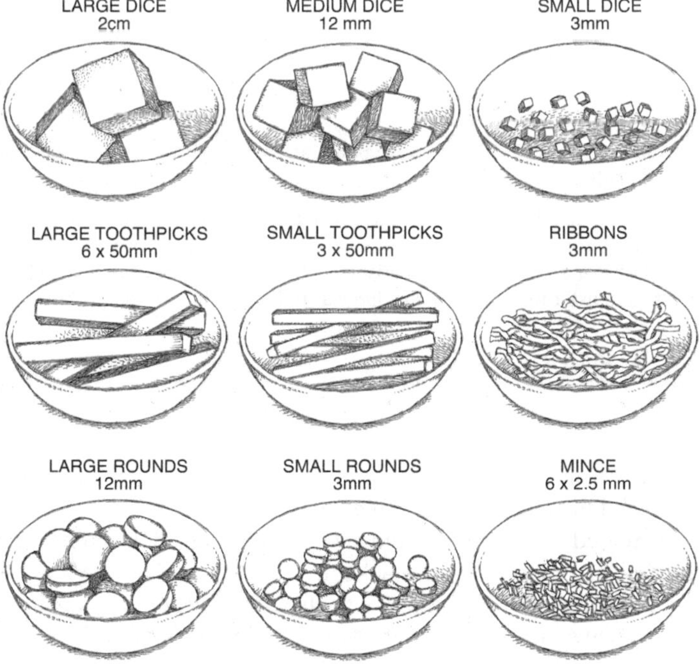

Ingredients

Various vegetables (e.g., carrots, cucumber, bell pepper, celery)

Procedure

Cut the vegetables into sticks or various shapes for dipping. Serve with dips of choice (see Dips section for some options).

Cook's Notes: ______________________________

____

Cold Spicy Nuts

You store these nuts in the freezer. It's the contrast between the cold and the warming spices that makes them really work. You also get a firmer texture, and they last longer. You can make this recipe with any mixed nut blend, but I'm partial to 2 parts peanuts, 1 part walnuts, and 1 part almonds. I always have a big freezer bag full of these ready for any occasion.

Ingredients
4 large egg whites
2 kg (5 lbs) mixed nuts (suggested: 2 parts peanuts, 1 part walnuts, 1 part almonds)
125 mL (½ cup) granulated sugar
60 mL (¼ cup) light brown sugar
20 mL (4 tsp) ground cinnamon
30 mL (2 tbsp) cayenne pepper
25 mL (1 ½ tbsp) salt
5 mL (1 tsp) cumin

Cooking Instructions
Preheat the oven to 163°C (325°F). In a large bowl, whisk the egg whites until they form soft peaks, then fold in all spices and sugars.

In another bowl, combine the spiced egg whites with the nuts, ensuring an even coating. Spread the mixture evenly across parchment-lined or silicone-mat-lined sheet pans. Bake for 30 min, stirring at the 20-min mark to ensure even crisping. Remove from the oven, let cool to room temperature, then transfer to the freezer in a large airtight container or freezer bag.

Assembly/Presentation
Serve directly from the freezer for the best cold-spicy contrast. Serve in a shallow dish.

Taste: Sweetness from sugar, saltiness, and spiciness from cayenne and cinnamon.
Texture: Crisp and firm, enhanced by freezing.
Temperature: Served cold, with warming spices creating a delayed heat effect.
Tactility: Eaten by hand, offering a snackable crunch.
Temptation: The caramelized coating contrasts with the cool, firm texture.
Trigeminal: The cayenne and cinnamon activate a tingling, warming sensation.

Alternations
Adjust spice level: Reduce cayenne for less heat or increase for more intensity.
Vary texture: Try cashews or pecans for different shapes and crunch.

Cook's Notes: ____
____

Crispy Chickpeas with Spices

Looking for a healthy, crunchy alternative to potato chips? These golden, protein-rich chickpeas offer a satisfying crunch, bold seasoning, and endless flavor possibilities. Enjoy them as a snack or a crunchy topping for salads and soups. Keep a batch on hand for whenever you need something crispy and flavorful.

Ingredients
1 can (400 g/14 oz) chickpeas, drained and rinsed
15 mL (1 tbsp) olive oil
5 mL (1 tsp) paprika
2.5 mL (½ tsp) cumin
2.5 mL (½ tsp) chili powder
2.5 mL (½ tsp) salt

Cooking Instructions
Preheat the oven to 200 °C (400 °F). Pat the chickpeas dry with a clean tea towel to remove excess moisture—this helps them crisp up.

In a bowl, toss the chickpeas with olive oil, paprika, cumin, chili powder, and salt, ensuring an even coating. Spread them evenly on a parchment-lined baking sheet. Bake for 20–30 min, stirring halfway through, until golden and crispy.

Let cool before serving to enhance the crispiness.

Assembly/Presentation
Serve cold alongside crackers, toasted bread, or raw vegetables. They also make a great crunchy topping for soups and salads.

Taste: Mild saltiness with a subtle warmth from the spices.
Texture: Crispy on the outside, slightly firm inside. Becomes crunchier as it cools.
Temperature: Can be served at room temperature or chilled for extra crunch.
Tactility: Small and easy to pick up, with a dry, crisp bite.
Temptation: Golden brown chickpeas coated in vibrant spices make them visually appealing.
Trigeminal: The chili powder provides a gentle heat that lingers slightly.

Alternations
Adjust spice level: Add more chili powder for extra heat, or reduce it for a milder snack.
Vary texture: Extend baking time by 5–10 min for a firmer crunch.

Cook's Notes: ______________________________
____

Stuffed Dates with Goat Cheese and Walnuts

This simple yet luxurious snack combines the natural sweetness of dates, the creamy richness of goat cheese, and the satisfying crunch of walnuts. With just a minute of prep time, these make for a quick appetizer or an elegant finger food that looks as good as it tastes.

Ingredients
12 large dates, pitted
100 g (3.5 oz) soft goat cheese
12 walnut halves

Cooking Instructions
Make a small incision in each date and remove the stone. Spoon or pipe the goat cheese into the cavity of each date, ensuring an even fill. Press a walnut half into the cheese, slightly embedding it for stability.

Assembly/Presentation
Serve immediately or store in the refrigerator until ready to enjoy. Arrange on a small serving plate for a refined look.

Taste: Natural sweetness of dates with cheese's saltiness and bitterness from the walnuts.
Texture: Soft and creamy inside, with a firm bite from the date and a crunchy walnut topping.
Temperature: Best served cool or at room temperature.
Tactility: Small, smooth, and easy to pick up; slightly sticky exterior with a creamy center.
Temptation: Glossy date exterior with a contrast of creamy white cheese and golden walnuts.
Trigeminal: Low levels but you may consider adding spices to the goat cheese.

Alternations
Add more contrast: Drizzle lightly with honey for extra sweetness or a few drops of balsamic glaze for tang.
Change creaminess: Try blue cheese, mascarpone, or ricotta.
Increase savoriness: Wrap in crispy prosciutto or bacon for a sweet-salty contrast.

Cook's Notes: ______________________________
____

Mini Pizzette

A cross between pizza and bruschetta, these mini pizzette are a quick and delicious way to use up leftovers or explore new flavor combinations. The crispy base, melty cheese, and variety of toppings make each bite an exciting multisensory experience.

Ingredients
4 slices whole meal bread or pita bread
60 mL (4 tbsp) tomato sauce
100 g (3.5 oz) grated mozzarella cheese

Various toppings: (e.g., sliced tomato, mushroom, chorizo, ham, olive, bell pepper, red onion, fresh basil)

Cooking Instructions
Preheat the oven to 200 °C (392 °F).

Spread the tomato sauce evenly over the bread or pita, ensuring full coverage to prevent dryness. Sprinkle the grated mozzarella on top, distributing it evenly. Arrange your chosen toppings to create a variety of textures and flavors. Place on a baking sheet and bake for 10–15 min, or until the cheese is melted and golden.

Assembly/Presentation
Serve warm. Cut into smaller pieces for easy sharing or enjoy whole for a satisfying bite.

Taste: Sweet and umami-rich tomato sauce and caramelized cheese.
Texture: Crispy and crunchy with soft cheese and toppings to provide contrast.
Temperature: Best served hot from the oven.
Tactility: Hand-held, with a firm but tender base and stretchy melted cheese.
Temptation: A vibrant mix of colors from toppings and a golden layer of bubbling cheese.
Trigeminal: Chili flakes or spicy sausage provides a mild, lingering warmth.

Alternations
Change the base: Use pesto or a white sauce on top of other breads (e.g., naan)
Increase pungency: Drizzle with garlic-infused olive oil post-baking.
Add tactile: Serve with dipping sauces such as tzatziki.

Cook's Notes: ____
____

Tangy Musubi

SPAM may not be the most elegant meat, but its processing for long shelf life gives it a unique melt-in-your-mouth texture. The textural quality of SPAM is enhanced by pan-frying and pairs well with the firm yet slightly sticky rice. A seasoning blend of rice vinegar, sugar, and salt, called sharizu, adds a tangy-sweet balance, making each bite more dynamic. Typically used in moderation to complement rather than overpower the dish, a slightly stronger dose of sharizu can help those with limited smell experience its full depth. The final wrap of nori seaweed provides a slight chew and crispness, completing this Hawaiian take on Japanese sushi.

Ingredients

1 can SPAM Lite (low sodium), sliced lengthwise into six sheets (½ inch thick)
250 mL (1 cup) premium sushi rice
375 mL (1 ½) cups water
80 mL (⅓ cup) rice vinegar
30 mL (2 tbsp) turbinado sugar
10 mL (2 tsp) salt
5 cm × 5 cm (2 × 2 inch) piece of kombu
15 mL (1 tbsp) wasabi
1 sheet roasted seaweed, sliced lengthwise into six long strips
15 mL (1 tbsp) neutral oil

Cooking Instructions

In a small saucepan over medium-low heat, combine rice vinegar, sugar, salt, and kombu. Stir until dissolved, then turn off heat and cover. This is the sharizu.

Rinse the rice in a mesh strainer under running water, moving it around until the water runs clear. In a stovetop pot, combine rinsed rice with water and bring to a boil. Cover, reduce heat to the lowest setting, and cook for 15 min. Alternatively, use a rice cooker with the recommended settings.

Once rice is done, fluff it with a fork to release steam. Transfer to a wide dish to cool. When still warm but not steaming, remove kombu and mix in 60 mL (4 tbsp) of sharizu. Cover and refrigerate for at least 1 h.

Remove rice from the fridge. Rub hands with some sharizu to prevent sticking. Divide into six equal portions, shaping each into a compact rectangle.

Heat a skillet with 1 tbsp oil over high heat. When the oil smokes, add three SPAM slices, searing for 1 minute per side. Repeat for remaining slices.

Assembly/Presentation

Place a rice rectangle in the center of a seaweed strip.
Spread ¼ tsp wasabi on top, then place a fried SPAM slice over it. Wet the inner edge of the seaweed strip and wrap it around the musubi, securing it like sushi.
Serve musubi warm or at room temperature. Pair with soy sauce and something pickled.

Taste: A balance of umami from SPAM and tangy-sweet rice sharizu.
Texture: Crispy edges of SPAM contrast with sticky, slightly firm rice, while nori provides a chewy wrap.
Temperature: Served with hot SPAM and cool rice, activating temperature-sensitive receptors.
Tactility: Handheld, wrapped in a slightly crisp seaweed sheet.
Temptation: The golden-brown SPAM, glossy rice, and dark nori strip create visual appeal.
Trigeminal: The wasabi introduces a sharp, sinus-clearing heat.

Alternations
Add sweetness: Brush SPAM with a teriyaki glaze before frying.
Increase heat: Add more wasabi or a drizzle of chili oil.
Make it crunchier: Roll rice in tempura flakes before wrapping.

Cook's Notes: ____
____

Olive-Spread Toast with Burrata

Vinegar is a living ingredient formed through fermentation, where yeasts convert carbohydrates into alcohol, then acetic acid, enhancing its acidic profile. While wine-based vinegars are common, any sugary liquid can undergo this transformation, producing unique acid profiles. This recipe layers acidity from vinegars, juices, pickles, and cheeses, engaging taste receptors independently of smell.

Texture is an essential part of the eating experience, and rheology is the study of it, or the science of how matter flows and deforms. Semi-liquid foods like spreads can change through heat or blending, altering their mouthfeel. This recipe creates a thick, coarse olive spread, but adjusting blending time or oil content can modify its texture. A longer blend results in a smoother consistency, while extra oil creates a thinner, smoother spread. Experimenting with these elements allows for a customized textural experience.

Ingredients

120 mL (½ cup) black olives, dried and pitted
3 cloves garlic, peeled
5 mL (1 tsp) Dijon mustard
5 caper berries, destemmed
1 bay leaf
2.5 mL (½ tsp) cayenne pepper, dry and ground
30 mL (2 tbsp) red wine vinegar
15 mL (1 tbsp) orange juice
1.25 g (¼ tsp) salt
60 mL (¼ cup) olive oil
4 slices whole grain bread
2 balls burrata
20 mL (4 tsp) honey

Cooking Instructions

In a food processor or blender, combine olives, garlic, mustard, caper berries, bay leaf, cayenne, vinegar, orange juice, and salt. Blend at a low speed for a few seconds, then slowly increase to medium speed over 30 s while drizzling in olive oil.

Generously butter both sides of the bread and toast until golden brown. Let cool slightly before spreading.

Assembly/Presentation

Spread a thick layer of olive mixture onto each slice of toast. Top each toast with a ball of burrata, carefully slicing down the middle to let it open up. Drizzle with honey.

Garnish with fresh herbs and a light sprinkle of flaky sea salt.

Taste: Briny olives, tangy vinegar, sweet honey, and creamy burrata.
Texture: The coarse olive spread contrasts with crispy toast and soft burrata.
Temperature: Best served at room temperature for optimal flavor and mouthfeel.
Tactility: Handheld or use utensils to enjoy each bite.
Temptation: Serve on a wooden board or rustic ceramic plate to complement the textures.
Trigeminal: Mile warming effect from the cayenne and olive oil.

Alternations

Make it smoother: Blend longer for a silkier consistency.
Enhance umami: Add anchovies or Parmesan to the olive spread.

Cook's Notes: ____
____

Fruit and Berry Skewers

Fruit skewers are a simple yet sensory-rich snack or dessert. They offer a vibrant mix of sweetness and acidity, complemented by varied textures and a cool, refreshing temperature. Their colorful presentation enhances visual appeal, while the ability to eat them with your fingers adds a playful and interactive element to the experience.

Ingredients

Various fruits and berries (e.g., strawberries, blueberries, grapes, melon, pineapple, kiwi)

Wooden skewers

Optional Enhancements:
Coarse chili powder
Lime juice
Ginger syrup
A pinch of salt

Cooking Instructions

Wash all fruits thoroughly. Cut larger fruits such as melon, pineapple, and kiwi into bite-sized pieces (see chopping chart under Veggie sticks), keeping smaller berries whole.

Assembly/Presentation

Thread the fruit onto wooden skewers, alternating colors, and textures to create variety.

Serve immediately or refrigerate until ready to enjoy for an extra cooling effect. Arrange on a platter, fanning the skewers out to highlight the vibrant colors.

Taste: Most fruits naturally provide sweetness with contrast from acidic fruits such as pineapple, kiwi, and different berries (e.g., currants, gooseberries).
Texture: Juicy, soft, and chewy contrasts.
Temperature: Best served chilled, making it ideal for hot days or as a light dessert.
Tactility: Handheld and easy to grip, with varied firmness between fruits.
Temptation: The natural rainbow of fruit that alternates in color.
Trigeminal: Chili powder provides a subtle heat, and ginger syrup adds a gentle warmth.

Alternations

Citrus burst: Add orange or tangerine segments for extra brightness.
Sweeten it up: Serve with honey-yogurt dip or drizzle with dark chocolate.
Herbal cooling: Sprinkle with mint leaves for freshness.
Try different fruit combinations: Try mango, dragon fruit, or passion fruit.

Cook's Notes: ____

____

Mains and Sides

Spaghetti a la Bolognese

Prep Time: [30 min] | *Cook Time*: [50 min] | *Serves*: [4]

Spaghetti Bolognese is a classic dish in most kitchens: with its umami, richness, and varied consistency, the dish speaks directly to our pleasure centers. Serve a fresh bitter salad with a nice vinaigrette to provide a crunchy and bitter counterpoint to the softness of the dish—add more texture and tactility by making crispy pesto bread, which is eaten with your fingers.

Sensory Profile (based on 6 Ts)

TASTE ■□□□□ *Sweet* ■■■□□ *Salty* ■■□□□ *Bitter* ■■■■□ *Sour* ■■■■■ *Umami*

TEXTURE ■■■□□ *Chunky* ■■■□□ *Crunchy* ■□□□□ *Chewy*

TEMP *Cold Salad* | *Hot dish* *TRIGEMINAL* ■□□

TACTILE *Hand-held bread* | *Utensils* *TEMPTATION* ■■□□□

Ingredients

Bolognese sauce

400 g (14 oz) ground beef, 10–12 % fat
1 large onion, finely chopped (about 200 g/7 oz)
2 garlic cloves, crushed (about 6 g/0.2 oz)
15 mL (1 tbsp) tomato paste
5 mL (1 tsp) paprika
30–45 mL (2–3 tbsp) tamari or soy sauce
1 can (400 g/14 oz) chopped tomatoes
15 mL (1 tbsp) balsamic vinegar
5 mL (1 tsp) sugar
5 mL (1 tsp) fresh thyme leaves, finely chopped
15 mL (1 tbsp) olive oil

Salad

Half head frisée or romaine lettuce (about 150 g/5 oz)
15 mL (1 tbsp) Dijon mustard
30 mL (2 tbsp) vinegar
45 mL (3 tbsp) olive oil
4 cherry tomatoes (about 60 g/2 oz), mixed colors
Coarse salt and pepper

Pesto Bread

8 slices bread (about 240 g/8 oz)
30 mL (2 tbsp) pesto

Accompaniments

450 g (1 lb) spaghetti
Fresh Parmesan for serving

Cooking Instructions

Preheat the oven to 180°C (360°F).

Heat a pot over medium heat and sauté onions, garlic, and thyme in olive oil for 2–3 min. Add ground beef and cook until browned. Stir in tomato paste, paprika, tamari, and balsamic vinegar; continue to sauté for 1–2 min. Add chopped tomatoes, bring to a simmer, and cook for at least 30 min (longer for richer flavor). Season with sugar, salt, and pepper to taste.

Wash and drain the lettuce in a strainer. Heat a pan on high heat, coat cherry tomatoes with a little olive oil, and sauté until browned. Whisk together Dijon mustard, vinegar, and olive oil for the vinaigrette. Season with salt & pepper to taste.

Slice bread into 1 cm thick slices and spread a thin layer of pesto over the top. Bake for 10 min or until crispy and golden.

Bring a large pot of salted water to a rolling boil. Add spaghetti and cook al dente (as per package instructions). Drain and toss with a drizzle of olive oil to prevent sticking.

Assembly/Presentation

Plate the spaghetti and top with Bolognese sauce, and garnish with freshly grated parmesan and basil leaves.

Place crispy pesto bread on the side of dish alongside bitter salad dressed in mustard vinaigrette.

Taste: Umami from beef and tamari; sour notes from tomato and vinegar.
Texture: Chunky sauce, chewy pasta, crunchy salad, and bread.
Temperature: Hot pasta and bread paired with a cold salad.
Trigeminal: Mild warmth from paprika and mustard.
Tactile: Pasta and salad with utensils; bread is hand-held.
Temptation: Moderate visual contrast of red sauce, green leaves, and golden bread.

Alterations

Vegetarian option: Swap ground beef for lentils or plant-based mince to maintain texture.
More creaminess: Stir a dollop of mascarpone into the sauce for extra creaminess.
Added texture: Toss salad with toasted walnuts for added crunch.

Cook's Notes: ____
____

Avocado Toast with Fried Egg

Prep Time: [15 min] | *Cook Time*: [15 min] | *Serves*: [2]

On those days when you need to move quickly, you don't need to compromise on the 6 T's, this is a quick and easy multisensory dish. There is temperature contrast in the form of cold avocado and hot fried egg, as well as texture variety with pickled cornichons and toasted bread. The dish is given an extra kick with lemon juice and Tabasco. Enjoy.

Sensory Profile (based on 6 Ts)

TASTE ■■□□□ *Sweet* ■■■□□ *Salty* ■□□□□ *Bitter* ■■■□□ *Sour* ■■■■□ *Umami*
TEXTURE ■■■■■ *Creamy* ■■■■□ *Crunchy* ■■■□□ *Velvety*
TEMP ❄ *Cold accompaniments* | 🔥 *Hot egg* *TRIGEMINAL* 🌶🌶□
TACTILE ✋ *Hand-held bread* *TEMPTATION* ■■■□□

Ingredients

Toast and eggs

2 slices whole-grain bread (80 g/2.8 oz each)
2 ripe avocados (approx. 400 g/14 oz total)
15 mL (1 tbsp) lemon or lime juice
30 g (1 oz) cream cheese
2 large eggs (100 g/3.5 oz total)
10 g (2 tsp) butter, for frying
2.5 mL (½ tsp) Tabasco or sambal oelek
30 g (1 oz) cornichons or sliced pickles
Coarse salt and freshly ground pepper, to taste

Cooking Instructions

Toast bread until golden and let cool slightly. Spread each slice with cream cheese.

Prep the avocado: Halve, pit, and slice the avocados. Drizzle with lemon juice to prevent browning.

Heat butter in a pan over medium heat. Crack the eggs, season with salt and pepper, and fry until whites are set but yolks remain runny.

Assemble/Presentation

Arrange avocado slices over the cream cheese. Place a fried egg on each slice. Drizzle with Tabasco and scatter pickles on top.

Taste: Umami from egg yolk and cream cheese; sour from lemon and pickles.
Texture: Creamy avocado and cheese, crispy toast, and pickles.
Temperature: Cold avocado and pickles topped with a hot, runny egg.
Trigeminal: Tabasco adds a gentle, lingering heat.
Tactile: Entirely hand-held for a playful eating experience.
Temptation: Bright green avocado and golden egg yolk offer moderate visual appeal.

Alterations
Add Protein: Top with smoked salmon or crispy bacon for extra umami.
Extra Heat: Drizzle additional hot sauce or sprinkle chili flakes.
Egg Cooking Variations: Try a poached or soft-boiled egg for a different texture.

Cook's Notes: ____
____

Buttered Cauliflower with Hazelnuts, Chimichurri, and Feta Spread

Prep Time: [20 min] | *Cook Time*: [40 min] | *Serves*: [4]

The best way to serve a whole cauliflower is to bake it with plenty of butter. In this recipe, the cauliflower is balanced by the Argentine spicy sauce chimichurri, which has a nice freshness, acidity, and chili heat. The toasted hazelnuts add some crunchiness, and with the cool feta spread, we get a satisfying, temperature contrast, saltiness, and umami in the dish.

Sensory Profile (based on 6 Ts)

TASTE ■■□□□ *Sweet* ■■■□□ *Salty* ■□□□□ *Bitter* ■■■□□ *Sour* ■■□□□ *Umami*
TEXTURE ■■■■■ *Soft* ■■□□□ *Crunchy* ■■■□□ *Tender*
TEMP ❄ *Cold spread* | 🔥 *Hot dish* *TRIGEMINAL* 🌶□□
TACTILE 🍴 *Utensils* *TEMPTATION* ■■■□□

Ingredients

Buttered Cauliflower
1 head cauliflower (about 900 g/2 lb)
100 g (⅓ cup) soft butter
Juice of 1 lemon (30 mL/2 tbsp)

Chimichurri
45 mL (3 tbsp) olive oil
50 mL (3 tbsp plus 1 tsp) red wine vinegar
15 mL (1 tbsp) chopped pickled jalapeños
3 scallions, chopped (about 45 g/1.6 oz)
1 garlic clove, chopped (about 5 g/0.2 oz)
15 mL (1 tbsp) chopped mixed herbs (parsley, oregano, cilantro)
Salt and pepper

Feta Spread
100 g (3.5 oz) feta
55 mL (¼ cup) Greek yogurt
Zest of ½ lemon

Accompaniments
30 g (1 oz) hazelnuts, toasted

Cooking Instructions

Preheat the oven to 210 °C (410 °F). Trim cauliflower leaves and stem so it sits flat. Rub with melted butter mixed with lemon juice, season with salt, and roast for 25–30 min, basting once, until deeply golden.

Whisk olive oil, vinegar, jalapeños, scallions, garlic, and herbs in a bowl; season and let rest. In a dry pan, toast hazelnuts over medium heat until fragrant.

Blend feta, yogurt, and lemon zest in a food processor until smooth and creamy.

Assembly/Presentation

Cut roasted cauliflower into wedges. Arrange on a platter, drizzle with any pooled butter-lemon sauce, spoon chimichurri over the top, and sprinkle with hazelnuts. Serve immediately with feta spread on the side.

Taste: Sour and umami stand out from chimichurri vinegar and feta.
Texture: Soft cauliflower, crunchy hazelnuts, creamy spread.
Temperature: Hot roasted cauliflower contrasts with cold spread.
Trigeminal: Mild heat from jalapeños in the chimichurri.
Tactile: Fork-friendly wedges and dollops of spread.
Temptation: Golden-brown surface, green herbs, and white spread give visual appeal.

Alterations
Add Bolder Flavors: Roast the garlic before blending it into the feta spread.
Cauliflower Alternative: Try this with broccoli or Romanesco for a bitter flavor profile.

Cook's Notes: ____
____

Meatballs in Curry with Spicy Rice

Prep Time: [60 min] | *Cook Time*: [60 min] | *Serves*: [4]

Meatballs in curry are a fantastic dish, especially when you add a little flair to it and remember all the elements to achieve the perfect balance between basic flavors and texture. We make sure to sauté the curry well with apples and onions and add sweet and sour elements in the form of a chutney. It's also important to have some crunch and freshness, for which we have chosen peanuts and mango: all of these elements work very well together with the soft meatballs. Add a warm naan bread on the side to engage your tactile senses and add another texture to the meal.

Sensory Profile (based on 6 Ts)

TASTE ■■■■□ *Sweet* ■■■□□ *Salty* ■□□□□ *Bitter* ■■□□□ *Sour* ■■■■■ *Umami*

TEXTURE ■■■■□ *Creamy* ■■■■□ *Tender* ■■■□□ *Crunchy*

TEMP *Hot dish* *TRIGEMINAL* ■■□

TACTILE *Hand-held* | *Utensils* *TEMPTATION* ■■■□□

Ingredients

Meatballs

400 g (14 oz) ground beef & pork
1 onion, finely grated (≈150 g)
2 eggs
20 g (½ oz) oats
100 mL (½ cup) milk
Salt & pepper to taste

Curry Sauce

15 mL (1 tbsp) curry powder
300 mL (1¼ cups) chicken stock
2 onions, chopped (≈300 g)
3 apples, diced (≈300 g)
1 tbsp (15 mL) curry powder
200 mL (¾ cup) heavy cream
2 tbsp (30 mL) spicy mango chutney
1 tbsp (15 mL) lemon juice
1 tbsp (15 mL) tamari or soy sauce

Rice

2 cups (370 g) basmati rice
4 cups (960 mL) water
1 cinnamon stick
3 cardamom pods
4 cloves
1/2 tsp salt

Accompaniments

1 mango, diced (≈200 g)
A handful of peanuts (≈30 g/1 oz)
4 naan breads
Fresh cilantro (optional)

Cooking Instructions

In a bowl, combine the grated onion with the ground meat, curry powder, salt, and pepper. Stir in the eggs, oats, and milk until just combined, then cover and chill in the refrigerator for 30 min.

While the meatball mixture chills, bring the chicken stock to a gentle simmer. Use a spoon to shape the meat mixture into balls and slide them carefully into the hot stock. Poach for 5–7 min until just firm, then transfer the meatballs to a plate and strain the stock, returning a reduced amount to the pot.

In the same pot, sauté the chopped onions and diced apples over medium heat until softened. Stir in the curry powder and toast for 30 s, then pour in the reduced stock and heavy cream. Let the sauce simmer for 5–8 min before seasoning with tamari, lemon juice, mango chutney, salt, and pepper. Return the meatballs to the sauce and heat through for 2–3 min.

To cook the rice, bring the water, cinnamon stick, cardamom pods, cloves, salt, and a pinch of salt to a boil. Add the basmati rice, cover, and reduce heat to low; simmer for 10 min. Turn off the heat and let the rice rest, covered, for 7 min. Remove the spices and fluff with a fork.

Assembly/Presentation

Spoon spiced rice into bowls, ladle meatballs and curry sauce on top, scatter diced mango and peanuts over each, and serve with warm naan.

Taste: Apple and mango chutney deliver a bright sweet-sour lift.
Texture: Creamy sauce contrasts tender meatballs and crunchy peanuts.
Temperature: Served hot to keep spices aromatic and sauce smooth.
Trigeminal: Curry powder and chutney provide a warm, lingering heat.
Tactile: Naan invites hand-held scooping for multisensory engagement.
Temptation: The golden curry flecked with red chutney and bright mango is visually enticing.

Alterations

Herbal boost: Stir in chopped cilantro or mint just before serving for freshness.
Heat level: Add sliced green chilies or a pinch of cayenne to the sauce for extra kick.
Dairy-free: Swap heavy cream for coconut milk to maintain creaminess without dairy.

Cook's Notes: ____
____

Indian Daal with Spiced Rice

Prep Time: [40 min] | *Cook Time*: [60 min] | *Serves*: [4]

Daal is an amazing dish that embodies great vegetarian food: it is so finely balanced and has lots of vegetable umami flavor, so you don't even notice there's no meat in it. The red lentils give a texture like ground beef, the chili and ginger add some spiciness, the coconut milk adds richness, and the tamari provides a deep umami flavor. On top of the dish, the roasted chickpeas provide a chewy texture, while the cold raita provides a temperature counterbalance.

Sensory Profile (based on 6 Ts)

TASTE ■■■□□ *Sweet* ■■■□□ *Salty* ■□□□□ *Bitter* ■■□□□ *Sour* ■■□□□ *Umami*

TEXTURE ■■■□□ *Creamy* ■■□□□ *Crunchy* ■■□□□ *Chewy*

TEMP *Cold topping* | *Hot dish*

TRIGEMINAL ■□□

TACTILE *Hand-held papadums* | *Utensils*

TEMPTATION ■■■□□

Ingredients

Daal

200 g (7 oz) red lentils
2 large onions, chopped
5 carrots, grated
1 large knob of ginger, peeled & chopped
≈30 g (1–2) red chili peppers, chopped
3 cloves garlic, minced
3 stalks lemongrass, chopped
3–4 leaves kaffir lime leaves, optional
1 lime, zest & juice
400 g (14 oz) chopped tomatoes
400 mL (14 fl oz) coconut milk
15 mL (1 tbsp) sugar
45 mL (3 tbsp) tamari (or soy sauce)

Rice

370 g (2 cups) basmati rice
960 mL (4 cups) water
1 cinnamon stick
3 cardamom pods
4 cloves
2.5 mL (½ tsp) salt

Papadums

4 papadums
Oil, for frying

Roasted Chickpeas

100 g (3.5 oz) chickpeas
30 mL (2 tbsp) lemon juice
15 mL (1 tbsp) curry powder
Oil, for frying
Salt & pepper, to taste

Raita

200 g (7 oz) Greek yogurt (10% milk fat)
A handful mint leaves (≈15 g), finely chopped

Accompaniments

Roasted coconut flakes (as desired)

Cooking Instructions

If using dried chickpeas, soak them overnight in cold water, then drain and boil in salted water for 30 min, reserving some of the cooking liquid to adjust the daal's consistency later. Meanwhile, blend chili peppers, garlic, lemongrass, and ginger into a coarse paste. In a large pot, heat a splash of oil and sauté the chopped onions until soft, then add the spice paste and toast for about one minute. Stir in the red lentils, kaffir lime leaves, tamari, chopped tomatoes, and coconut milk, bring to a boil, then reduce heat and simmer for 10 min. Add the grated carrots and continue cooking gently for 30–40 min, stirring occasionally and thinning with reserved chickpea water if needed. Finish by seasoning with sugar, lime zest & juice, salt, and pepper to taste.

While the daal simmers, bring the water with the cinnamon stick, cardamom pods, cloves, and salt to a boil. Stir in the rice, cover, reduce to low heat, and simmer for 10 min. Remove from heat, let rest covered for 7 min, then fluff with a fork.

Drain the cooked or canned chickpeas and pat them dry. Heat a little oil in a pan and fry until crisp, then toss with curry powder and lemon juice, cooking briefly until it sizzles. In a small bowl, mix the yogurt with chopped mint to make the raita. Finally, heat oil in a skillet and fry each papadum for a few seconds until puffed and crisp, turning once and draining on paper.

Assembly/Presentation

Ladle hot daal over spiced rice, top with roasted chickpeas, add a dollop of raita, and sprinkle with roasted coconut flakes. Serve alongside crisp papadums.

Taste: Lentils and tamari provide umami depth balanced by bright lime and subtle sweetness.
Texture: Creamy daal contrasts chewy chickpeas, crunchy papadums, and silky raita.
Temperature: Daal and chickpeas are served hot; raita cools the palate.
Tactile: Papadum scooping invites hand-held engagement; spoon works for raita.
Temptation: Deep-orange daal speckled with chickpeas and white raita is visually enticing.
Trigeminal: Chili and ginger deliver a gentle, lingering heat.

Alterations
More Heat: Add extra chili flakes or cayenne for more heat
Add texture: Stir coconut flakes into the daal before serving for richer coconut flavor.
Make it Smoother: Blend half of the daal for a creamier consistency.

Cook's Notes: ____
____

Plaice with Curry Remoulade and Beetroot Coleslaw

Prep Time: [30 min] | *Cook Time*: [40 min] | *Serves*: [4]

Cornmeal works brilliantly as a breadcrumb substitute, resulting in really crispy and delicious fried plaice fillets. It's quicker than the classic breading method with flour, egg, and breadcrumbs, and the cornmeal coating doesn't absorb as much oil. Always fry in a mixture of half butter and half olive oil to increase the cooking temperature of the butter and prevent it from burning too easily. For the classic fried fish, remoulade is necessary, and you can elevate the dish by making it from scratch. This recipe also includes a beetroot coleslaw, which provides a delightful crunchy contrast to the fish fillets and boiled potatoes with dill.

Sensory Profile (based on 6 Ts)

TASTE ■■■□□ *Sweet* ■■■□□ *Salty* ■□□□□ *Bitter* ■■□□□ *Sour* ■■□□□ *Umami*
TEXTURE ■■■■■ *Crispy* ■■■□□ *Creamy* ■■■■□ *Crunchy*
TEMP ❄ *Cold salad* | 🔥 *Hot dish* *TRIGEMINAL* 🌶□□
TACTILE 🍴 *Utensils* *TEMPTATION* ■■■□□

Ingredients

Plaice
4 plaice fillets, skinned
90 g (3¼ oz) polenta flour
Butter and oil for frying
Salt & pepper

Curry Remoulade
250 g (8¾ oz) pickles, chopped
200 g (7 oz) Greek yogurt
15 mL (1 tbsp) curry powder

Beetroot Coleslaw
300 g (10½ oz) peeled raw beetroot, julienned
½ punnet red grapes
1 lime, zest & juice
5 mL (1 tsp) sugar
A handful (≈30 g/1 oz) walnuts, toasted

Accompaniments
800 g (28 oz) potatoes
1 bunch dill, chopped
½ lemon, in wedges
Salt & pepper

Cooking Instructions

Peel and julienne the beetroot, toss with sugar, lime zest, & juice, then fold in grapes and toasted walnuts. In a small bowl, mix the pickles with yogurt and curry powder to make the remoulade.

Boil the potatoes in salted water until tender, drain, chop into bite-sized pieces, and toss with chopped dill, salt, and pepper.

Pat the plaice dry, season, and coat both sides in polenta flour. Heat equal parts butter and oil in a skillet over medium-high. Fry the fillets 3–4 min per side until golden and crisp, then transfer to a warm plate.

Assembly/Presentation

Arrange potatoes and beetroot coleslaw beside the plaice, add a generous dollop of curry remoulade, and garnish with lemon wedges.

Taste: The creamy remoulade and sweet beetroot slaw balance the savory, crispy fish.
Texture: Crunchy slaw and creamy sauce contrast the ultra-crispy coating.
Temperature: Fish and potatoes served hot; slaw and remoulade cool the palate.
Tactile: Utensils manage the remoulade, while lemon-squeezing adds interactivity.
Temptation: Golden fillets against bright red slaw and green dill entice at first glance.
Trigeminal: Curry powder in the remoulade offers a gentle warmth.

Alterations

Enhance crispiness: Let coated fillets rest 5–10 min before frying.
Boost freshness: Stir chopped parsley or chives into the remoulade.
Grill citrus: Char lemon wedges briefly for smoky brightness before serving.

Cook's Notes: ____
____

Danish "Frikadeller" with Potato Salad

Prep Time: [40 min] | *Cook Time*: [60 min] | *Serves*: [4]

If you have not had Danish "frikadeller" before, you're in for a treat: it's a classic dish in Denmark and typically each family has their own way to make it. In these ones we have added hazelnuts and cranberries to the meatball mixture to give a little more texture for the mouth. In the potato salad, there's some smoked cheese and honey to add a smoky flavor and some sweetness; there are also radishes to give a crispy and bitter contrast. If you're not a fan of smoked cheese, use Greek yogurt instead.

Sensory Profile (based on 6 Ts)

TASTE ■■■□□ *Sweet* ■■■□□ *Salty* ■□□□□ *Bitter* ■■■□□ *Sour* ■■■■□ *Umami*
TEXTURE ■■■■■ *Crispness* ■■■□□ *Crunchy* ■■■■□ *Tender*
TEMP *Cold salad* | *Hot dish* *TRIGEMINAL* ■□□
TACTILE *Utensils* *TEMPTATION* ■■■■□

Ingredients

Frikadeller

500 g (1.1 lb) ground veal & pork
1 onion, finely chopped
2 eggs
35 g (1¼ oz) oats
200 mL (¾ cup + 2 tbsp) milk
1 handful chopped hazelnuts
1 handful dried cranberries
2 tsp salt
Butter & oil for frying
Pepper to taste

Potato Salad

750 g (1.65 lb) potatoes
185 g (6½ oz) smoked cheese spread
100 g (3½ oz) 18% crème fraîche
15 mL (1 tbsp) Dijon mustard
15 mL (1 tbsp) honey
15 mL (1 tbsp) lemon juice
50 g (1¾ oz) pickles, chopped
1 bunch radishes, thinly sliced
1 bunch parsley, chopped
Salt & pepper to taste

Accompaniments

Rye bread croutons (from Crunch Kit)

Cooking Instructions

In a bowl, mix the ground veal and pork with onion, eggs, oats, milk, hazelnuts, cranberries, salt, and pepper, then cover and chill for 30 min. While it rests, boil the potatoes in salted water until tender, drain, cool slightly, and slice; whip the smoked cheese with crème fraîche, mustard, honey, and pickles, then toss with the potatoes and season to taste.

Heat butter and oil in a skillet over medium heat, shape the chilled meat mixture into patties (dipping the spoon in fat to prevent sticking), and fry until deeply golden and cooked through, about 4–5 min per side. Meanwhile, trim and soak the radish slices in iced water for extra crispness.

Assembly/Presentation

Arrange the potato salad on a platter, place the hot frikadeller alongside, scatter the radishes and parsley over the salad, and serve with rye bread croutons.

Taste: The sweet cranberries and honey counterbalance savory, umami-rich meat, and cheese.
Texture: Ultra-crispy frikadeller and croutons contrast tender meat and creamy salad.
Temperature: Meatballs served hot; salad and radishes refresh the palate.
Tactile: Utensils for salad and bread for scooping make it interactive.
Temptation: Golden patties against bright red radishes and green parsley are visually enticing.
Trigeminal: A hint of black pepper adds a pungent sting.

Alterations

Toast nuts: Pre-toast hazelnuts at 160 °C (320 °F) for 10 min before chopping for added crunch.
Lighten dressing: Swap smoked cheese spread for Greek yogurt to reduce richness.
Increase appeal: Serve on a rustic wooden board with a side of crunchy croutons for texture contrast.

Cook's Notes: ____
____

Quesadillas with Chicken and Tomato Salad

Prep Time: [30 min] | *Cook Time*: [40 min] | *Serves*: [4]

Mexican quesadillas are a great dish for busy days, and you can fill them with almost anything. They create good textural contrasts between the crispy surface and the soft filling. The salsa provides a little chili heat, and the cold creme adds freshness and coolness to the dish. In this recipe, there is a tomato salad, but you can play around with this as you like. Try it with cooked beans, cactus, or whatever you feel like.

Sensory Profile (based on 6 Ts)

TASTE ■■■□□ *Sweet* ■■■■□ *Salty* ■□□□□ *Bitter* ■■■□□ *Sour* ■■■□□ *Umami*

TEXTURE ■■■■□ *Crispiness* ■■■□□ *Soft* ■■■□□ *Juicy*

TEMP ❄ *Cold salad* | 🔥 *Hot dish* *TRIGEMINAL* 🌶🌶□

TACTILE ✋ *Hand-held* *TEMPTATION* ■■■□□

Ingredients

Quesadillas

8 large flour tortillas
330 g (12 oz) chicken breast, sliced
2 garlic cloves, thinly sliced
¼ white onion, finely chopped
5 mL (1 tsp) smoked paprika
2.5 mL (½ tsp) cumin
100 g (3½ oz) corn kernels
100 g (3½ oz) grated cheddar cheese
3 tomatoes, sliced
1.25 mL (¼ tsp) Mexican hot sauce (e.g., Cholula)
30 mL (2 tbsp) oil or melted lard for frying

Tomato Salad

400 g (14 oz) mixed cherry tomatoes, halved
1 red onion, thinly sliced
15–30 mL (1–2 tbsp) sliced jalapeños
10 mL (2 tsp) apple cider vinegar
1 lime, zest & juice
15 mL (1 tbsp) extra virgin olive oil

Accompaniments

200 g (7 oz) spicy tomato salsa
100 g (3½ oz) crème fraîche (18% milk fat) or Mexican crema
Nacho chips
Salt & pepper, to taste

Cooking Instructions

Heat oil in a skillet over medium-high and sauté garlic and white onion until fragrant. Add chicken strips, smoked paprika, cumin, salt, and pepper, and cook until golden and cooked through. Toss in corn and Tabasco, then transfer to a bowl. In that same skillet, warm the tortillas one at a time, piling half the cheese, chicken mixture, tomato slices, and a sprinkle of cheese on one, topping with a second tortilla; press gently and cook 2–3 min per side until the cheese melts and the tortillas are golden. Keep warm in a low oven while you repeat.

Meanwhile, in a bowl combine halved cherry tomatoes, red onion, jalapeños, lime zest, & juice; season with salt to make the tomato salad. Keep cold until served.

Assembly/Presentation

Serve quesadillas hot with bowls of spicy salsa and crema (or crème fraiche) alongside the tomato salad and fried corn tortilla chips for scooping.

Taste: Smoky paprika and Tabasco heat are balanced by sweet tomatoes and creamy cheese.
Texture: Ultra-crispy tortillas contrast the creamy filling and juicy salad.
Temperature: Quesadillas served hot; salad and crème fraîche cool the palate.
Tactile: Hand-held wedges and chips invite interactive eating.
Temptation: Golden tortillas against bright red salsa and salad create strong visual appeal.
Trigeminal: Jalapeños and Tabasco deliver a lingering, tingly warmth.

Alterations

Alter heat: Dried pepper has a variety of temporal heat profiles; try different ones at your local Mexican grocer.
Add texture: Sprinkle toasted pumpkin seeds inside before folding.
Customize protein: Swap chicken for sautéed mushrooms or black beans.

Cook's Notes: ____

____

Beet Tartare

Prep Time: [30 min] | *Cook Time*: [300 min] | *Serves*: [4]

This vegetarian version of tartare uses the same flavorings and gets plenty of umami flavor from roasted beets. When vegetarian food is done right, you won't miss the meat, and that's certainly the case with this delightful beet tartare, which has a good balance of flavors, a kick from mustard, and a nice crunch from peanuts and fried capers.

Sensory Profile (based on 6 Ts)

TASTE ■■□□□ *Sweet* ■■■□□ *Salty* ■□□□□ *Bitter* ■■□□□ *Sour* ■■■□□ *Umami*
TEXTURE ■■■□□ *Course* ■■■□□ *Moist* ■■■■□ *Soft*
TEMP ❄ Cold tartare | 🌡 Warm bread TRIGEMINAL ■□□
TACTILE ✋ Hand-held bread | 🍴 Utensils TEMPTATION ■■■■□

Ingredients

Beet Tartare

700 g (1.5 lb) whole beets1 handful capers
30 mL (2 tbsp) oil for frying
1 handful dried cranberries, chopped
1 handful salted peanuts
1 small bunch dill, chopped
30 mL (2 tbsp) lemon juice
10 mL (2 tsp) Dijon or whole-grain mustard
15 mL (1 tbsp) extra-virgin olive oil
Salt & pepper to taste

Accompaniments

Toasted bread

Cooking Instructions

Place the peeled beets in an ovenproof dish, toss with a little oil, salt, and pepper, and roast at 100 °C (210 °F) for about 3 h until very tender. Let cool to room temperature. Meanwhile, pat capers dry and fry in hot oil until golden and crispy; drain on paper.

Cut the cooled beets into chunks and pulse in a food processor (or grate by hand) until a coarse, smooth texture forms. Transfer to a bowl and stir in mustard, olive oil, lemon juice, cranberries, peanuts, and chopped dill. Season with salt and pepper.

Assembly/Presentation

Mold the tartare onto plates or toasts using a ring or spoon. Top with fried capers and extra dill sprigs. Serve alongside warm toasted bread and a simple green salad.

Taste: Roasted beets and mustard give earthy-sweet depth, balanced by lemon's brightness and salty capers.
Texture: Smooth beet purée contrasts the crisp capers and chewy cranberries.
Temperature: Tartare is served cold; bread adds a warm counterpoint.
Tactile: Scoop with bread for hand-held bites or use utensils for a composed starter.
Temptation: The deep-red tartare dotted with golden capers and green dill is highly appealing.
Trigeminal: A hint of mustard delivers a gentle, tangy bite.

Alterations

Add color: Scatter pomegranate seeds for jewel-like brightness.
Swap nuts: Use toasted pistachios or sunflower seeds for a different crunch.
Add creaminess: Top with whipped goat cheese for extra richness.

Cook's Notes: ______________________________
____

Portobello Carpaccio with Garlic Crostini

Prep Time: [25 min] | *Cook Time*: [30 min] | *Serves*: [4]

A classic carpaccio consists of a thinly sliced beef fillet marinated with lemon juice and olive oil. In this recipe, the meat is replaced with portobello mushrooms, which, like the beef, offer plenty of umami flavor. The preparation is the same, and it gets an extra umami kick from freshly shaved Parmesan, acidity and bitterness from the vinaigrette, and a little crunch from the toasted pine nuts and garlic crostini.

Sensory Profile (based on 6 Ts)

TASTE ■□□□□ *Sweet* ■■■□□ *Salty* ■■□□□ *Bitter* ■■■■□ *Sour* ■■■■□ *Umami*
TEXTURE ■■■■□ *Crispy* ■■■□□ *Tender* ■■■□□ *Oily*
TEMP ❄ *Cold dish* | 🌡 *Warm bread*
TACTILE ✋ *Hand-held bread* | 🍴 *Utensils*
TRIGEMINAL 🌶□□
TEMPTATION ■■■■□

Ingredients

Carpaccio

4 portobello mushrooms
20 g (0.7 oz) Parmesan cheese, shaved
45 mL (3 tbsp) lemon juice
45 mL (3 tbsp) extra-virgin olive oil
5 mL (1 tsp) honey
½ bunch arugula, washed
10 mL (2 tsp) Dijon mustard
30 g (1 oz) pine nuts, toasted
1 bunch chives, chopped
Salt & pepper, to taste

Garlic Crostini

1 small baguette, sliced
2 garlic cloves, halved
Oil for brushing
Salt & pepper, to taste

Cooking Instructions

Whisk together lemon juice, olive oil, honey, mustard, salt, and pepper. Slice mushrooms as thin as possible and arrange them overlapping on a chilled platter; spoon half the vinaigrette over them and let sit for 5–10 min.

Toast pine nuts in a dry pan until golden. Brush baguette slices with oil and toast under a broiler or in a 180 °C (355 °F) oven until crisp; rub each with the cut side of a garlic clove and season lightly with a pinch of salt.

Assembly/Presentation

Drizzle the remaining vinaigrette over the mushrooms, then scatter shaved Parmesan, chopped chives, arugula, and pine nuts on top. Serve immediately with warm garlic crostini alongside.

Taste: Bright lemon and honey balance the savory mushrooms and salty Parmesan.
Texture: Silky mushroom slices contrast the ultra-crisp crostini and pine nuts.
Temperature: Mushrooms served cold; crostini provide a warm counterpoint.
Tactile: Use utensils for composed bites or crostini for hand-held dipping.
Temptation: Glossy mushrooms, snowy cheese shavings, and golden crostini create strong visual appeal.
Trigeminal: A hint of mustard gives a gentle, tangy bite.

Alterations

Add richness: Drizzle with aged balsamic reduction for sweet-tart depth.
Add freshness: Scatter fresh basil or mint leaves for an herbal lift.
Add texture: Sprinkle panko crumbs toasted in butter for extra crunch. Reference crunch kit for other options.

Cook's Notes: ____

____

Jambalaya with Trinity Pickles

Prep Time: [30 min] | *Cook Time*: [90 min] | *Serves*: [6]

Jambalaya is a dish built on layers of flavor, texture, and bold heat, combining deeply seasoned proteins, aromatic rice, and a balance of tangy, spicy, and umami-rich elements. The spiciness in this dish is pronounced, coming from a blend of cayenne, paprika, and chili powder that infuses the rice and proteins with a lingering warmth. Texture plays a vital role in shaping this dish as well. The pull of tender chicken, the snappiness of shrimp, and the crunch of pickles contrast with rice that is nicely coated in smooth, flavorful oil. These varying textures, combined with the dish's heat, create a complex mouthfeel that builds as you eat.

Sensory Profile (based on 6 Ts)

TASTE ■□□□□ *Sweet* ■■■□□ *Salty* ■■□□□ *Bitter* ■■■□□ *Sour* ■■■■□ *Umami*

TEXTURE ■■■■□ *Tender* ■■■■□ *Moist* ■■■□□ *Silky*

TEMP Optional cold pickles | Hot dish TRIGEMINAL 🌶🌶🌶

TACTILE Utensils TEMPTATION ■■□□□

Ingredients

Jambalaya

225 g (1.5 cups) trinity pickles, strained
1 L (4 cups) chicken stock
285 g (1.5 cups) long grain rice
22.5 g (1.5 tbsp) tomato paste
1 jalapeño, capped and seeded
3 garlic cloves, finely chopped
30 g (2 tbsp) spice mix *(see note)*
15 g (1 tbsp) butter
15 mL (1 tbsp) oil
7.5 g (1.5 tsp) dried oregano
500 g (1 lb) chicken thighs
225 g (½ lb) andouille sausage, sliced
225 g (½ lb) shrimp, unpeeled
4–6 sprigs of flat-leaf parsley, chopped
1 lemon, quartered

Cajun Spice Mix

30 g (2 tbsp) cayenne pepper
30 g (2 tbsp) paprika
15 g (1 tbsp) white pepper
15 g (1 tbsp) black pepper
60 g (4 tbsp) chili powder
15 g (1 tbsp) garlic powder

Trinity Pickles

180 mL (¾ cup) white wine vinegar
60 g (4 tbsp) sugar
22.5 g (1.5 tbsp) salt
1 green bell pepper, diced to ½-inch cubes
1 onion, diced to ½-inch cubes
4 celery stalks, diced to ½-inch cubes
2 bay leaves
4–6 sprigs of flat-leaf parsley
2 tsp yellow mustard seeds
Juice of 1 lemon

Cooking Instructions

Cajun Spice Mix

Combine all ingredients and store in an airtight container.

Trinity Pickles

Evenly distribute diced bell pepper, onion, and celery into sterilized 16 oz mason jars. Add 1 bay leaf, 2–3 parsley sprigs, 1 tsp mustard seeds, and juice of ½ lemon to each jar.

In a saucepan, bring 360 mL (1.5 cups) water to a boil. Remove from heat and stir in vinegar, sugar, and salt until dissolved.

Pour hot brine over vegetables in each jar, ensuring they are fully submerged. Seal jars, let cool to room temperature, then refrigerate. Allow at least 24 h for flavors to develop.

Chicken Stock

Heat 45 mL (3 tbsp) canola oil in a pressure cooker until smoking. Sear 1 whole chicken back and 2 wings until deeply browned. Add ½ onion (quartered) and vegetable trimmings from trinity pickles. Pressure cook for 45 min, then allow pressure to release naturally (~30 min). Alternatively, simmer on the stovetop for 2.5 h.

Jambalaya

Heavily salt both sides of the chicken thighs and let them sit overnight in the fridge.

Blend together 750 mL (3 cups) of chicken stock, tomato paste, jalapeño, garlic, 5 g (1 tsp) of salt, and 15 g (1 tbsp) of spice mix for 30 s. In a pot, combine blended mixture with rice, 3 bay leaves, and 5 g (1 tsp oregano). Bring to a boil, then cover and set to the lowest heat setting. Cook undisturbed for 15 min.

Turn on the oven broiler. Heat butter and oil in a large oven-proof pan on high heat until slightly smoking. Add chicken and sear both sides for 2 min. Add sausage, shrimp, remaining 250 mL (1 cup) of chicken stock, and 15 g (1 tbsp) spice mix. Stir to incorporate, then slide into the oven under the broiler for 7 min, then cool undisturbed for 5 min.

Remove rice from heat, fluff, and transfer to a large bowl. Add protein mixture (including its liquid) and trinity pickles (not including pickling liquid). Stir until combined, then cover with a towel and let the flavors meld for 10 min.

Assembly/Presentation

Serve scoops of jambalaya on a plate and garnish with flat-leaf parsley and a lemon wedge. Serve hot.

Taste: A balance of salty, umami, and heat, with sour brightness from pickles and lemon.
Texture: Moist rice, tender proteins, and chewy sausage add variety.
Temperature: Served hot, with cooling contrast from pickles. Dish tastes great cold as well.
Tactility: Eaten with a fork, but scoopable with chips or can be wrapped in lettuce.
Temptation: The deep-red spice-infused rice, golden-brown proteins, and bright green parsley. *Trigeminal*: Spice mix contributes a lingering warmth.

Alterations

Add more heat: Increase cayenne in spice mix or add fresh other peppers.
Lessen the heat: Reduce cayenne and replace jalapeno with poblano.

Cook's Notes: ____

____

Smash Burger with Pungent Mustard

Prep Time: [20 min] | *Cook Time*: [10 min] | *Serves*: [4]

A great smash burger delivers a sensory contrast—an ultra-crispy, caramelized crust balanced by a tender, juicy center. However, the high heat and smashing technique can cause fat loss, reducing flavor and moisture. To counter this, we incorporate gelatin, a structural protein derived from collagen, to help retain fat and enhance juiciness. This method is similar to the technique used in Chinese soup dumplings, where gelatin melts during cooking, creating a luscious, almost soupy texture. Here, it ensures that each bite of the burger remains succulent despite the intense sear.

The pungent mustard in this recipe provides a complementary sharpness to the rich beef. Mustard's chemesthetic effect comes from allyl isothiocyanate, a compound that activates TRPA1 receptors, creating a subtle heat and tingling sensation. This adds depth to the burger's flavor profile, enhancing the interplay of richness, acidity, and spice for a more dynamic eating experience.

Sensory Profile (based on 6 Ts)

TASTE ■□□□□ *Sweet* ■■■■□ *Salty* ■■□□□ *Bitter* ■■□□□ *Sour* ■■■■■ *Umami*

TEXTURE ■■■■□ *Juicy* ■■■□□ *Fatty* ■■■■■ *Crispy*

TEMP *Hot dish*

TRIGEMINAL ■■□

TACTILE *Hand-held*

TEMPTATION ■■□□□

Ingredients

Burgers

450 g (1 lb) ground beef, 80:20 protein-to-fat ratio
1 egg yolk
15 mL (1 tbsp) olive oil
7.2 g (0.25 oz) unflavored powdered gelatin
5 g (1 tsp) salt
2.5 g (½ tsp) black pepper
½ head iceberg lettuce, shredded
4 slices (120 g) cheddar cheese
1 onion, chopped
8–10 pickle slices (slightly sweet version)
4 burger buns
60 g (4 tbsp) mayonnaise

Pungent Mustard

180 mL (¾ cup) white wine
180 mL (¾ cup) apple cider vinegar
2 garlic cloves, peeled
60 g (½ cup) yellow mustard seeds
10 g (1 tbsp + 2 tsp) black pepper
5 g (1 tsp) salt
5 g (1 tsp) sugar
15 mL (1 tbsp) high-quality olive oil

Cooking Instructions

Mustard

In a blender, combine white wine, apple cider vinegar, garlic, mustard seeds, black pepper, salt, and sugar. Cover and refrigerate overnight.

Add olive oil and blend on high for 1 minute. Store in an airtight container in the fridge for up to 2 months.

Burgers

Mix salt, pepper, and gelatin in a bowl. Loosen the ground beef with two forks, then sprinkle in the dry mixture. Add the egg yolk and gently mix with forks. Cover and refrigerate if making ahead.

Divide the meat into four equal portions. Squeeze each portion five times between your hands (creaming technique) and roll into balls.

Preheat a cast-iron skillet over high heat (outdoor grills preferred for smoky flavor). Place one or two meatballs onto the skillet and smash them into thin patties (6 mm/¼-inch thickness) using a flat spatula. Cook for 2 min. Meanwhile, toast the buns on the grill, rotating every 30 s. Flip patties, top with cheese, and cook for 30 more seconds. Remove from heat along with the buns.

Assembly/Presentation

On the bottom bun, spread mayonnaise, add chopped onions, and place the patty on top. Next, add pickles and shredded lettuce. Spread pungent mustard on the remaining bun and place it on top. Serve immediately with fries or a side of your choice.

Taste: The burger is salty, tangy, and umami-rich, with mild bitterness from mustard seeds and black pepper.

Texture: A crispy crust contrasts with the juicy beef, while lettuce and pickles add crunch.

Temperature: Served hot, fresh from the skillet.

Tactility: Eaten by hand, with a soft bun and crisp toppings.

Temptation: It's a bit sloppy in appearance, but seeing a burger makes you salivate.

Trigeminal: The pungent mustard and black pepper introduce mild heat.

Alterations

Add more heat: Increase black pepper in the mustard or add chili flakes.

Add sweet flavor: Use honey mustard or peanut butter instead of mustard.

Cook's Notes: ____

____

Çılbır with Breadcrumbs

Prep Time: [10 min] | *Cook Time*: [10 min] | *Serves*: [2]

Çılbır combines creamy yogurt, softly poached eggs, and a warm chili-infused butter for a dish that contrasts temperatures and textures. The acidity of vinegar in the poaching water helps adjust the pH, allowing the egg whites to coagulate more efficiently, resulting in a delicate yet structured poached egg. Served hot, with a pungent yogurt base and a drizzle of spiced butter, this dish highlights the interplay of richness, tang, and heat.

Sensory Profile (based on 6 Ts)

TASTE ■□□□□ *Sweet* ■■■□□ *Salty* ■■□□□ *Bitter* ■■■□□ *Sour* ■■■□□ *Umami*
TEXTURE ■■■■□ *Creamy* ■■■□□ *Soft* ■■□□□ *Crunchy*
TEMP *Room Temp* *TRIGEMINAL* ■■□
TACTILE *Hand-held bread* | *Utensils* *TEMPTATION* ■■■■□

Ingredients

2 eggs
15 mL (1 tbsp) vinegar
250 mL (1 cups) whole-milk yogurt
1 garlic clove, finely grated
5 g (1 tsp) salt
30 g (2 tbsp) butter
5 ml (1 tsp) olive oil
½ tsp red pepper flakes (Aleppo pepper recommended)
125 mL (½ cup) mint leaves, chopped

Cooking Instructions

One day prior to cooking, in a bowl, mix yogurt with grated garlic and salt. Let it sit in the fridge overnight to infuse. Prior to cooking, remove from the fridge to come to room temperature.

Bring a pot of salted water to a gentle simmer (not a rolling boil). Add vinegar to the water—this lowers the pH, helping the egg whites firm up faster for a smoother poached egg. Crack each egg into a small bowl, create a vortex in the water by making a circular motion with a spoon, and gently slide in the eggs. Poach for 3–4 min, then remove with a slotted spoon and place in a bowl with lukewarm water to wash off the vinegar.

Prepare the Chili Butter: In a small pan, melt butter with olive oil over medium heat. Add red pepper flakes and stir until fragrant, about 30 s. Remove from heat.

Assembly/Presentation

Spread the yogurt mixture on a plate or shallow bowl. Place poached eggs on top of the yogurt and drizzle the warm chili butter over the eggs. Sprinkle on mint leaves. Serve with warmed pita bread.

Taste: The yogurt provides a tangy base and butter adds richness.
Texture: Yogurt and butter add creaminess, with a soft bite from the poached eggs.
Temperature: Served warm, with a contrast between the hot eggs and cool yogurt.
Tactility: Eaten with a spoon or scooped up with bread.
Temptation: The golden chili butter contrasts against the white yogurt and soft poached egg.
Trigeminal: The chili flakes introduce mild heat, while mint provides cooling.

Alterations
Add more texture: *Top with any items in the crunch kit.*
Add more color contrast: Top with green herbs (e.g., parsley).

Cook's Notes: ____
____

Lentil Casserole with Coconut Milk

Prep Time: [10 min] | *Cook Time*: [25 min] | *Serves*: [4]

This lentil casserole blends warm spices, creamy coconut milk, and fresh spinach for a comforting dish that is rich in umami and spice. The red lentils provide a velvety texture, while the coconut milk adds a silky richness. The gentle heat from chili powder balances the natural sweetness of tomatoes and coconut, making this dish a deeply satisfying sensory experience.

Sensory Profile (based on 6 Ts)

TASTE ■■□□□ *Sweet* ■■□□□ *Salty* ■■□□□ *Bitter* ■■■□□ *Sour* ■■■□□ *Umami*
TEXTURE ■■■■□ *Velvety* ■■■□□ *Creamy* ■■□□□ *Chewy*
TEMP *Hot dish*
TACTILE *Hand-held bread* | *Utensils*
TRIGEMINAL ■□□
TEMPTATION ■■□□□

Ingredients

Casserole
15 mL (1 tbsp) olive oil
1 onion, chopped
2 garlic cloves, minced
5 mL (1 tsp) ground cumin
5 mL (1 tsp) ground coriander
2.5 mL (½ tsp) turmeric
1.25 mL (¼ tsp) chili powder (adjust to taste)
1 can (400 g/14 oz) chopped tomatoes
1 can (400 mL/14 fl oz) coconut milk
200 g (1 cup) red lentils, rinsed
400 mL (1⅔ cups) vegetable broth
100 g (3.5 oz) fresh spinach
Salt & pepper, to taste

Garnish
Fresh coriander
Red chili flakes

Cooking Instructions

In a large saucepan, heat the olive oil over medium heat. Sauté the onion and garlic until soft and translucent, about 5 min. Stir in the cumin, coriander, turmeric, and chili powder and cook for 1 min until fragrant. Add the chopped tomatoes, coconut milk, red lentils, and vegetable broth, bring to a simmer, and cook, stirring occasionally, until the lentils are tender but still hold their shape, about 20 min.

Stir in the spinach and cook just until wilted. Season with salt and pepper to taste.

Assembly/Presentation

Ladle into bowls, garnish with fresh coriander and red chili flakes, and serve warm with rice, naan bread, or quinoa.

Taste: Coconut milk and tomatoes create a sweet-tangy base enhanced by warm spices.
Texture: Creamy lentils and silky coconut milk contrast tender wilted spinach.
Temperature: Served warm to highlight its comforting richness.
Trigeminal: A touch of chili powder provides a gentle, lingering heat.
Tactile: Best enjoyed with a spoon or scooped up with crusty bread.
Temptation: The golden-orange bake dotted with bright green spinach is visually inviting.

Alterations

Add texture: Sprinkle toasted pumpkin seeds or chopped nuts for crunch.
Boost umami: Stir in a dash of soy sauce or miso for deeper savory notes.
Brighten flavor: Squeeze fresh lime juice over each serving for a zesty lift.

Cook's Notes: ______________________________
____

Baked Sweet Potato with Filling

Prep Time: [10 min] | *Cook Time*: [60 min] | *Serves*: [4]

This dish balances natural sweetness, creamy textures, and a touch of heat for a vibrant and satisfying meal. The caramelized baked sweet potatoes provide a soft, almost buttery base, while the black bean and corn filling adds freshness and a mild umami depth. Crunchy roasted seeds enhance the texture, making each bite a delightful contrast of smooth and crisp.

Sensory Profile (based on 6 Ts)

TASTE ■■■■□ *Sweet* ■■■□□ *Salty* ■■□□□ *Bitter* ■■□□□ *Sour* ■■□□□ *Umami*
TEXTURE ■■■■□ *Silky* ■■■■□ *Thick* ■■□□□ *Crunchy*
TEMP *Hot dish* *TRIGEMINAL* ■□□
TACTILE *Utensils* *TEMPTATION* ■■■■□

Ingredients
4 large sweet potatoes

Filling
1 can (400 g/14 oz) black beans, drained and rinsed
1 can (300 g/10½ oz) corn kernels, drained
1 avocado, diced
Juice of 1 lime
1 handful fresh coriander, chopped
Salt & pepper, to taste

Garnish
Roasted pumpkin seeds or pine nuts
½ tsp chili flakes (optional)
Additional fresh coriander

Cooking Instructions
Preheat the oven to 200 °C (400 °F). Pierce the sweet potatoes all over with a fork and place them directly on the oven rack. Bake until very tender when pierced, about 1 h. While they roast, stir together black beans, corn, avocado, lime juice, chopped coriander, salt, and pepper in a bowl. Add chili flakes if you would like a touch of heat.

Assembly/Presentation
Slice each baked potato lengthwise and gently mash the flesh to create a well. Spoon in the black bean mixture, then sprinkle with roasted pumpkin seeds and extra coriander. Serve immediately while warm.

Taste: Natural sweetness from the potato contrasts with tangy lime and savory beans.
Texture: Velvety potato flesh meets juicy corn and crunchy seeds.
Temperature: Served warm to highlight comforting flavors.
Trigeminal: Optional chili flakes provide a gentle warmth.
Tactile: Pick up to take a bite or eat filling with a spoon.
Temptation: Sweet potatoes, when stuffed, give off a variety of fall colors.

Alterations
Add texture: Stir in diced red bell pepper or cucumber for fresh crunch.
Boost creaminess: Dollop Greek yogurt or sour cream on top before serving.
Brighten flavor: Mix chopped tomatoes and a splash of vinegar into the filling.

Cook's Notes: ____
____

Quinoa Salad with Grilled Vegetables

Prep Time: [15 min] | *Cook Time*: [20 min] | *Serves*: [4]

A vibrant and nutritious salad that combines nutty quinoa, smoky grilled vegetables, and a tangy balsamic dressing. The mix of textures, from light and fluffy quinoa to tender grilled vegetables and crisp fresh herbs, makes this dish both satisfying and refreshing. Serve it warm for a comforting meal or chilled for a light, refreshing dish.

Sensory Profile (based on 6 Ts)

TASTE ■■□□□ *Sweet* ■■□□□ *Salty* ■■□□□ *Bitter* ■■■□□ *Sour* ■■□□□ *Umami*
TEXTURE ■■■■□ *Light* ■■■□□ *Crunchy* ■■□□□ *Moist*
TEMP *Room Temp*
TRIGEMINAL ■□□
TACTILE *Utensils*
TEMPTATION ■■■□□

Ingredients

For the Salad

185 g (1 cup) quinoa
480 mL (2 cups) vegetable broth
1 red pepper, cut into strips
1 yellow pepper, cut into strips
1 zucchini, sliced
1 red onion, cut into wedges
15 mL (1 tbsp) olive oil
Salt & pepper to taste

For the Dressing

45 mL (3 tbsp) olive oil
15 mL (1 tbsp) balsamic vinegar
5 mL (1 tsp) Dijon mustard
1 garlic clove, pressed
Salt & pepper to taste

Accompaniments

Pinch of chili flakes or black pepper (optional)
Fresh parsley or arugula

Cooking Instructions

Rinse the quinoa under cold water. In a saucepan, bring quinoa and vegetable broth to a boil, then reduce heat, cover, and simmer until tender and liquid is absorbed, about 15 min. Fluff with a fork and let cool slightly. Meanwhile, toss the peppers, zucchini, and onion with olive oil, salt, and pepper; grill over medium-high until tender with char marks, about 5–7 min per side.

Whisk together the dressing ingredients—olive oil, balsamic, mustard, garlic, salt, pepper, and chili flakes if using—adjusting seasoning to taste. In a large bowl, combine the quinoa and grilled vegetables, pour over the dressing, and toss until evenly coated.

Assembly/Presentation

Transfer the dressed salad to a serving platter or bowl, garnish with parsley or arugula, and serve warm or chilled.

Taste: The tangy balsamic and sweet peppers balance the nutty quinoa.
Texture: Light, fluffy quinoa pairs with juicy grilled veggies and crisp herbs.
Temperature: Serve warm to enhance aroma or chilled for a refreshing bite.
Trigeminal: A pinch of chili in the dressing adds a gentle warmth.
Tactile: Best eaten with a fork, stacking all components into a composed, flavorful bite.
Temptation: The colorful mix of red, yellow, and green vegetables is visually enticing.

Alterations

Seasonal Swap: Use eggplant, asparagus, or cherry tomatoes depending on availability.
Add creaminess: Crumble feta or goat cheese on top.
Boost brininess: Stir in chopped olives or capers for extra depth.

Cook's Notes: ______________________________
____

Vegetarian Burger with Crispy Salad and Homemade Fries

Prep Time: [20 min] | *Cook Time*: [30 min] | *Serves*: [4]

A satisfying and colorful vegetarian burger featuring hearty bean-quinoa patties, fresh crispy salad, and golden homemade fries. The contrast between the soft bun, crunchy lettuce, and crispy fries enhances the eating experience, while the seasoning in the patties provides layers of umami and warmth.

Sensory Profile (based on 6 Ts)

TASTE ■■□□□ *Sweet* ■■■■□ *Salty* ■□□□□ *Bitter* ■■□□□ *Sour* ■■■□□ *Umami*
TEXTURE ■■■□□ *Soft* ■■■□□ *Crispy* ■■□□□ *Firm*
TEMP ❄ *Cold salad* | 🔥 *Hot dish* *TRIGEMINAL* 🌶□□
TACTILE ✋ *Hand-held* | 🍴 *Utensils* *TEMPTATION* ■■■■□

Ingredients

Vegetarian Patties

1 can (400 g/14 oz) kidney beans, drained & rinsed
250 mL (1 cup) cooked quinoa
125 mL (½ cup) rolled oats
1 onion, finely chopped
2 garlic cloves, pressed
1 egg
60 mL (¼ cup) breadcrumbs
5 mL (1 tsp) ground cumin
2.5 mL (½ tsp) chili powder (adjust to taste)
Salt & pepper, to taste

Fries

4 large potatoes, peeled & cut into sticks
30 mL (2 tbsp) olive oil
Salt & pepper, to taste
½ tsp baking soda
5 mL (1 tsp) salt (for soaking)

For Serving

4 burger buns, toasted
Lettuce leaves
Tomato slices
Red onion rings
1 avocado, sliced
Your choice of dip (vegan mayo, ketchup, mustard)
Optional: sliced jalapeños

Cooking Instructions

Preheat the oven to 200 °C (400 °F). Soak the potato sticks in a bowl of cold water with the baking soda and 5 mL salt for 1 h. Drain thoroughly, pat dry, toss with olive oil, salt, and pepper, and spread in a single layer on a parchment-lined baking sheet. Bake 20–30 min, flipping once, until golden and crisp.

While the potatoes soak, mash the beans in a large bowl. Stir in quinoa, oats, onion, garlic, egg, breadcrumbs, cumin, chili powder, salt, and pepper until a sticky mixture forms. Shape into four patties. Heat a little oil in a skillet over medium heat and cook the patties 4–5 min per side, until firm and golden.

Assembly/Presentation

Lightly toast the burger buns. Layer lettuce, a patty, tomato, onion, and avocado on the bottom bun, add your dip, then top with the other half. Serve immediately with the fries and optional jalapeños.

Taste: Umami from the bean–quinoa patties and salty fries balanced by the sweet tomato and creamy avocado.

Texture: Soft, toasted buns and tender patties contrasted with crisp lettuce, juicy tomato, and crunchy fries.

Temperature: Warm patties, fries, and buns against the cool lettuce, tomato, and avocado.

Trigeminal: Chili powder in the patties and optional jalapeños deliver a gentle, tingling warmth.

Tactile: Hand-held assembly makes each bite interactive and satisfying.

Temptation: Layered colors of the burger create a visually inviting stack.

Alterations

Boost umami: Stir 1 tsp miso paste into the patty mixture for deeper savory notes.

Add acidity: Top burgers with quick-pickled red onions or a squeeze of lime for bright contrast.

Add richness: Spread guacamole or creamy hummus on the buns for an extra layer of indulgence.

Cook's Notes: ____

____

Dessert

Banana Split

Prep Time: [10 min] | *Cook Time*: [10 min] | *Serves*: [4]

The banana split is a classic on the menu and a nostalgic dessert. It has a nice balance between the basic tastes, sweetness and bitterness, and temperature opposites: the cold ice cream and the warm chocolate sauce. In this recipe, toasted coconut flakes add some extra crunch, but you can also use pistachios or peanuts as an alternative.

Sensory Profile (based on 6 Ts)

TASTE ■■■■■ *Sweet* ■■□□□ *Salty* ■■□□□ *Bitter* ■■□□□ *Sour* ■□□□□ *Umami*

TEXTURE ■■■■□ *Creamy* ■■■□□ *Chewy* ■■■□□ *Firm*

TEMP ❄ *Cold ice cream* | 🔥 *Hot sauce* *TRIGEMINAL* □□□

TACTILE 🍴 *Utensils* *TEMPTATION* ■■■■■

Ingredients

4 bananas
15 mL (1 tbsp) lemon juice
120 g (4 oz) vanilla ice cream (4 scoops of 30 g each)
300 mL (1¼ cups) whipping cream
150 g (5 oz) dark chocolate (70%), chopped
20 g (¾ oz) pistachios, toasted

Cooking Instructions

Toast the pistachios in a dry pan until golden and set aside. Warm 150 mL of the cream in a small saucepan over low heat, add the chopped chocolate, and stir until smooth; keep the sauce warm. Whip the remaining 150 mL cream until soft peaks form. Chill the ice cream in the fridge for 20 min to make scooping easier.

Assembly/Presentation

Peel and halve the bananas lengthwise, arrange in dishes, and drizzle with lemon juice. Place one scoop of ice cream between each banana half, spoon over warm chocolate sauce, add a dollop of whipped cream, and sprinkle with pistachios.

Serve immediately.

Taste: Sweet banana and ice cream offset by bittersweet chocolate.
Texture: Creamy ice cream and whipped cream, tender banana, crunchy nuts.
Temperature: Cold ice cream meets warm sauce.
Trigeminal: -
Tactile: Scooped and eaten with a spoon.
Temptation: The contrast of white cream, dark sauce, and golden banana is visually irresistible and a reminder that you can always be a kid when it comes to desserts.

Alterations

Add fruit: Top with fresh berries or sliced kiwi for brightness.
Swap cream: Use coconut whip and dairy-free chocolate for a vegan version.
Boost flavor: Stir cinnamon or orange zest into the chocolate sauce for aromatic depth.
Even more toppings: Sprinkle with toasted coconut flakes or cocoa nibs for extra crunch.

Pancakes with Fruit and Syrup

Prep Time: [10 min] | *Cook Time*: [15 min] | *Serves*: [4]

Soft, fluffy pancakes served with fresh fruit and drizzled with syrup for a delicious and visually appealing dish. The combination of warm pancakes and cool, juicy fruit creates a delightful contrast in texture and temperature.

Sensory Profile (based on 6 Ts)

TASTE ■■■■□ *Sweet* ■■□□□ *Salty* □□□□□ *Bitter* ■■□□□ *Sour* ■□□□□ *Umami*
TEXTURE ■■■□□ *Soft* ■■■■□ *Sticky* ■■□□□ *Chewy*
TEMP ❄ *Cold fruit* | 🔥 *Hot cakes* *TRIGEMINAL* □□□
TACTILE 🍴 *Utensils* *TEMPTATION* ■■■■■

Ingredients

Pancakes

200 mL (1 2/3 cups) all-purpose flour
5 mL (1 tsp) baking powder
1.25 mL (¼ tsp) salt
15 mL (1 tbsp) sugar
1 egg
300 mL (1 1/4 cups) milk
60 mL (4 tbsp) melted butter

For Serving

Fresh fruit (strawberries, blueberries, banana slices, apple wedges)
Maple syrup or other syrup

Optional

5 mL (1 tsp) grated fresh ginger in batter or syrup

Cooking Instructions

In a bowl whisk the flour, baking powder, salt, and sugar. In another bowl whisk the egg, milk, and melted butter, then pour into the dry ingredients and stir until just smooth. Let the batter rest 5 min.

Heat an additional knob of butter in a nonstick pan over medium. Pour in 60 mL (¼ cup) of batter for each pancake, tilting to spread evenly. Cook 1–2 min until bubbles form on top, flip, and cook another 1–2 min until golden. Stack on a warm plate while you finish the rest.

Assembly/Presentation

Stack 3–4 pancakes per plate, spoon the fresh fruit alongside or on top, and drizzle generously with syrup. Add a sprinkle of large crystal salt to enhance the entire flavor.

Taste: Sweet pancakes and syrup with a touch of salt.
Texture: Fluffy pancakes, juicy fruit, and sticky syrup.
Temperature: Warm pancakes meet cool fruit.
Trigeminal: -
Tactile: Eaten with a fork or hand-held when rolled.
Temptation: Golden stacks, colorful fruit, and glossy syrup look irresistible.

Alterations

Add spice: Stir ground cinnamon or nutmeg into the batter for warm aroma.
Boost richness: Fold mashed banana into the batter or add a swirl of cream cheese.
Consider pairing: This dish pairs well with hot bitter drinks like coffee or tea.

Cook's Notes: ____
____

Chocolate Mousse with Fresh Berries

Prep Time: [20 min] | *Cook Time*: [120 min] | *Serves*: [4]

A rich and airy chocolate mousse with a velvety texture, balanced by the freshness of juicy berries. The contrast of bitter dark chocolate, subtle sweetness, and slight tartness from the fruit makes this dessert both indulgent and refreshing.

Sensory Profile (based on 6 Ts)

TASTE ■■■■■ *Sweet* ■□□□□ *Salty* ■■■□□ *Bitter* ■■□□□ *Sour* □□□□□ *Umami*
TEXTURE ■■■■□ *Airy* ■■■■□ *Creamy* ■□□□□ *Juicy*
TEMP ❄ *Cold*
TRIGEMINAL □□□
TACTILE 🍴 *Utensils*
TEMPTATION ■■■■□

Ingredients

Chocolate Mousse
200 g (7 oz) dark chocolate (≥70% cocoa), chopped
4 eggs, separated
30 mL (2 tbsp) sugar
1 tsp vanilla extract
200 mL (¾ cup + 2 tbsp) whipping cream

For Garnish
Fresh berries (strawberries, raspberries, blueberries)

Cooking Instructions

Melt the chocolate in a heatproof bowl set over simmering water, stirring until smooth. Remove from heat and let cool slightly. In a bowl, whisk the egg yolks with half the sugar and vanilla until pale and fluffy, then fold in the melted chocolate. In a separate bowl, whip the cream to soft peaks and gently fold into the chocolate mixture. In another clean bowl, whisk the egg whites with the remaining sugar to stiff peaks, then carefully fold them into the mousse to preserve the airiness.

Assembly/Presentation

Divide the mousse among serving glasses or bowls. Cover and refrigerate for at least 2 h, until set. Just before serving, top each portion with a mix of fresh berries.

Taste: Sweet sugar and dark chocolate balanced by tart berries.
Texture: Airy mousse contrasts juicy berries.
Temperature: Served chilled.
Trigeminal: -
Tactile: Spoonable, melt-in-the-mouth.
Temptation: Dark glossy mousse crowned with vibrant red and blue berries.

Alterations

Enhance aroma: Stir 1 tsp espresso powder into the melted chocolate for deeper flavor.
Add crunch: Sprinkle crushed pistachios or candied hazelnuts on top.
Swirl flavor: Drizzle raspberry coulis onto mousse.

Yogurt Parfait with Granola and Berries

Prep Time: [15 min] | *Cook Time*: [0 min] | *Serves*: [2]

A refreshing parfait with layers of creamy Greek yogurt, crunchy granola, and juicy berries. The combination of sweet, tangy, and nutty flavors makes for a balanced and satisfying treat.

Sensory Profile (based on 6 Ts)

TASTE ■■■■□ *Sweet* ■■■□□ *Salty* ■■□□□ *Bitter* ■■□□□ *Sour* ■□□□□ *Umami*
TEXTURE ■■■■□ *Creamy* ■■■■□ *Crunchy* ■■■□□ *Course*
TEMP ❄ *Cold*
TACTILE 🍴 *Utensils*
TRIGEMINAL □□□
TEMPTATION ■■■□□

Ingredients
200 mL (¾ cup + 2 tbsp) Greek yogurt (full-fat)
100 mL (½ cup) granola
100 mL (½ cup) mixed berries (strawberries, blueberries, raspberries)
15 mL (1 tbsp) honey (optional)

Cooking Instructions
In two serving glasses or bowls, layer 50 mL of yogurt, 50 mL of granola, and 50 mL of berries, then repeat to fill. Drizzle honey over the top layer if desired. Serve immediately while the granola stays crisp and the yogurt is cold.

Assembly/Presentation
Present the parfait in clear glasses to showcase the distinct layers of white, gold, and vibrant berry hues.

Taste: Sweet yogurt and honey balanced by the tartness of berries.
Texture: Creamy yogurt contrasts with crunchy granola and juicy berries.
Temperature: Cold contrasts between cooler berries and yogurt
Trigeminal: -
Tactile: Best enjoyed with a spoon to appreciate each layer.
Temptation: Colorful strata of white, amber, and jewel-toned berries are visually enticing.

Alterations
Boost richness: Swirl in a spoonful of mascarpone or whipped cream for extra indulgence.
Brighten flavor: Fold in lemon zest or grated ginger for a zesty lift.
Enhance crunch: Sprinkle almond slivers or toasted pumpkin seeds before serving.

Cook's Notes: ____
____

Fruit Salad with Mint and Lime

Prep Time: [10 min] | *Cook Time*: [10 min] | *Serves*: [4]

A refreshing and vibrant fruit salad with a zesty lime dressing, fragrant mint, and a mix of sweet and tangy fruits.

Sensory Profile (based on 6 Ts)

TASTE ■■■■□ *Sweet* ■■□□□ *Salty* ■■□□□ *Bitter* ■□□□□ *Sour* ■□□□□ *Umami*
TEXTURE ■■■■□ *Juicy* ■■■□□ *Hard* ■■□□□ *Sticky*
TEMP ❄ *Cold*
TRIGEMINAL ■□□
TACTILE 🍴 *Utensils*
TEMPTATION ■■■■□

Ingredients

Fruit Salad

2 apples, diced
2 pears, diced
1 handful grapes, halved
1 handful blueberries
1 handful fresh mint leaves, chopped
15 mL (1 tbsp) lime juice
Zest of 1 lime
15 mL (1 tbsp) honey or maple syrup (optional)

Optional Additions

A pinch of chili powder
Ginger syrup (to taste)

Cooking Instructions

In a large bowl, combine the diced apples, pears, grapes, and blueberries. Add the chopped mint, lime zest, and lime juice, tossing gently to coat. Drizzle honey over the fruit if using, season with a pinch of salt, and toss once more. Chill for at least 5 min before serving, or enjoy immediately.

Assembly/Presentation

Spoon into a clear bowl or individual glasses to showcase the colorful layers. Garnish with a sprig of mint.

Taste: Sweet fruit balanced by tart lime.
Texture: Juicy fruit, soft berries, and crunchy apple.
Temperature: Served chilled for a refreshing effect.
Trigeminal: Optional chili adds a gentle warmth.
Tactile: Best eaten with a spoon to enjoy each layer.
Temptation: Bright colors of fruit and mint are visually enticing.

Alterations

Add brightness: Stir in orange or grapefruit segments for extra citrus zing.
Add crunch: Sprinkle toasted coconut flakes or chopped pistachios on top.
Add richness: Swirl in a spoonful of mascarpone or coconut cream between layers.

Cook's Notes: ______________________________
____

Cocktails

Long Forgotten Friends

This cocktail is crafted to be enjoyed by everyone, regardless of their sense of smell. It demonstrates smell loss for those who have not experienced it by serving it with a nose clip (or simply pinching your nose). For those who cannot smell or choose to use the nose clip, you'll experience a balance of sweetness and acidity, enhanced by the effervescence of the soda that adds a pleasant tingling sensation. Remove the clip, and you'll be filled with the full sensory experience lost by those with smell impairment, the smoky essence reminiscent of a campfire intertwined with the floral notes of late spring.

Ingredients

45 mL (1.5 oz) Del Maguey Vida Mezcal
45 mL (1.5 oz) fresh, strained red grapefruit juice
15 mL (0.5 oz) St-Germain Elderflower Liqueur
30 mL (1 oz) Squirt soda

Instructions

In a cocktail shaker filled with ice, combine mezcal, grapefruit juice, and elderflower liqueur.

Shake for 10 s to chill and dilute, then strain into a highball glass filled with ice.

Top with Squirt soda and serve immediately.

Optional: Provide a nose clip for those who want to experience the contrast between taste and aroma.

G & T

This gin and tonic maximizes carbonation retention by minimizing agitation and keeping all ingredients as cold as possible. Carbonation stays better in colder liquids and is lost through agitation and nucleation sites (particles in the liquid). To preserve effervescence, the gin and the glass should be stored in the freezer, and tonic water should be kept in the coldest part of the fridge until just before use. Glass and cans are preferred over plastic, as plastic bottles lose carbonation faster.

Ingredients

60 mL (2 oz) freezer-stored gin of choice
105 mL (3.5 oz) chilled tonic water
15 mL (0.5 oz) fresh, strained lime juice
1 large ice cube

Instructions

Tilt a freezing-cold rocks glass at a 45-degree angle and add ingredients in this order:

1. Pour measured freezer-stored gin.
2. Pour tonic water gently to reduce bubble loss.
3. Add lime juice
4. Slowly slide in a large ice cube.

Optional: Garnish with a lime wheel.

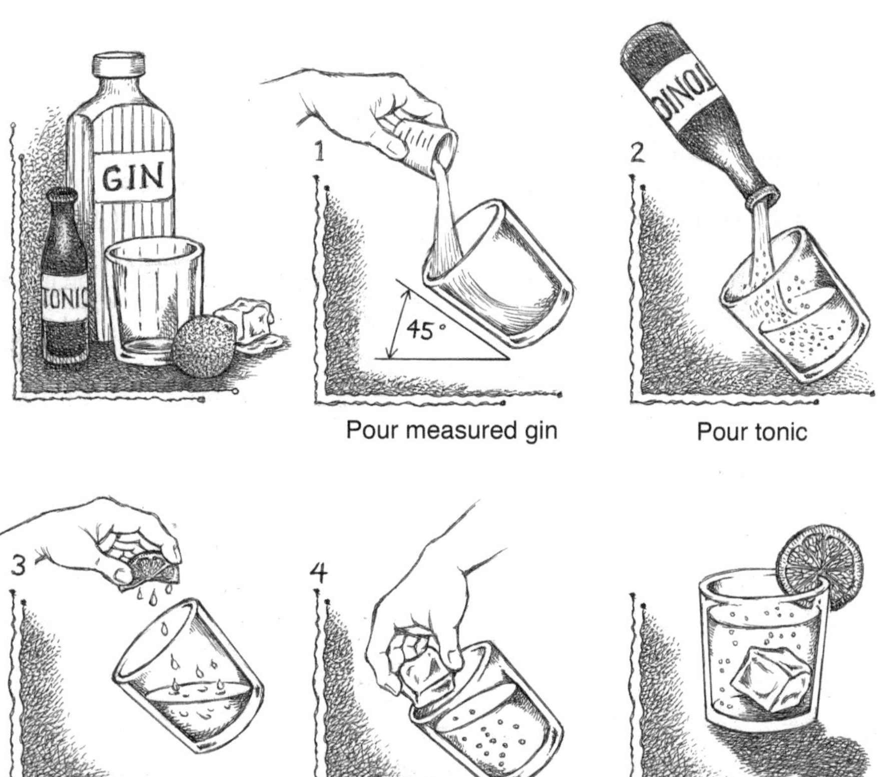

Cool Velvet

This cocktail highlights cooling as a sensory experience, making it particularly enjoyable for those with smell impairment. Cooling is a trigeminal response, where ingredients like mint activate receptors typically associated with cold, creating a refreshing sensation without lowering temperature. This drink pairs clear grappa's crisp bite with the herbal sweetness of Chambord and mint, while heavy cream smooths the texture. The addition of spearmint Altoids enhances the cooling effect, both through flavor and psychological color association. The pale color of the cocktail further reinforces the cooling perception, leveraging cross-modal effects where visual cues enhance cooling.

Ingredients

45 mL (1.5 oz) clear Grappa
15 mL (0.5 oz) Chambord
7.5 mL (0.25 oz) mint simple syrup (see note)
30 mL (1 oz) heavy cream
1 spearmint Altoids, crushed

Instructions

In a cocktail shaker filled with ice, combine grappa, Chambord, mint syrup, and heavy cream.

Shake for 10 s to chill, dilute, and emulsify.

Strain into a coupe glass.

Sprinkle a pinch of crushed spearmint Altoids on top before serving.

Mint Simple Syrup

In a saucepan, combine 250 g sugar and 250 g water, then add 80 mL (⅓ cup) dry mint and 4 spearmint Altoids. Bring to a boil, cover, and reduce heat, letting it sit for 2 min.

Remove from heat and steep for 1 h. Strain through cheesecloth or a fine mesh sieve to remove solids. Store in the fridge until ready to use.

The Big Green

The blender method is a powerful technique that not only extracts bold flavors but also enhances vibrant color in a drink. This cocktail is a striking bright green that holds its hue for at least half an hour, reinforcing the connection between sight and taste. We drink with our eyes first, and in this case, the lush green color sets the stage for the drink's star ingredient—flat-leaf parsley. Combined with gin, fresh lime juice, and a touch of heat from chili syrup, this cocktail is herbaceous, zesty, and subtly spicy.

Ingredients
60 mL (2 oz) Tanqueray gin
22 mL (0.75 oz) fresh lime juice
4 g flat-leaf parsley (leaves only)
15 mL (0.5 oz) chili simple syrup (see note)
Pinch of salt

Instructions
In a cocktail shaker (or blending container), combine gin, lime juice, and parsley. Blend using a hand blender or a regular blender, starting on low and increasing to high until smooth.

Transfer back to the shaker, then add chili simple syrup and a pinch of salt. Add ice and shake for 10 s to chill and dilute.

Double strain (using a cocktail strainer and fine-mesh sieve) into a coupe glass.

Chili Simple Syrup
In a saucepan, combine 250 g sugar and 250 g water, then add 4 crushed arbol chili peppers (seeds exposed). Bring to a boil, then cover and simmer for 2 min. Remove from heat and steep for 2 h. Strain through cheesecloth or a fine mesh sieve, then store in the fridge until use.

Bloody Buddy

A refined take on the Bloody Mary, this cocktail combines an umami-rich bloody mix with infused vodka, creating a deeply savory and layered drinking experience. The use of white soy sauce (or tamari and mirin) adds subtle saltiness and complexity, while pickled Japanese radish (takuan) and umeboshi plum provide a balanced acidity and a visually striking garnish. The horseradish and celery root-infused vodka enhances the drink's depth, making it an ideal choice for those who appreciate bold, savory flavors.

Ingredients

60 mL (2 oz) infused vodka (see note)
105 mL (3.5 oz) bloody mix (see note)
30 mL (1 oz) fresh lemon juice
15 mL (0.5 oz) white soy sauce (or 7.5 mL tamari + 7.5 mL mirin)
1 Japanese pickled radish (takuan)
1 Japanese pickled plum (umeboshi)

Instructions

Fill a highball glass with ice. Add infused vodka, bloody mix, lemon juice, and white soy sauce. Stir with a bar spoon for 10 s to chill.

Garnish by placing a slice of the pickled radish in half perpendicular to its length into the drink with part exposed from top. Next lay across the glass a toothpick piercing a pickled plum through the top of one quarter.

Infused Vodka

Sterilize a jar with at least 1200 mL capacity. In a blender, combine 100 mL vodka, 50 g peeled and cubed horseradish, and 50 g peeled and cubed celery root (alternatively 30 g of celery and 30 g of parsnip). Blend on high for 30 s. Transfer to the sterilized jar, then add 10 g crushed white pepper and 9 g kombu. Pour in 900 mL vodka and let sit for 1 week. Strain into a sterile container and store in the fridge until use.

Bloody Mix

In a blender, combine 480 mL (2 cups) 100% tomato juice, 480 mL (2 cups) whole canned high-quality tomatoes, and 30 mL (2 tbsp) miso. Blend until smooth, transfer to a container, and store in the fridge until use.

A Very Lemony Sgroppino

This variation on the classic Italian sgroppino enhances its bright, citrus-forward character with an extra punch of lemon from fresh juice, lemon sherbet, and preserved lemon rind. The cold temperature and creaminess from the sherbet contrast with the warmth of vodka and prosecco's carbonation, creating a dynamic multisensory experience. The richness of the sherbet softens the acidity, while maraschino cherry syrup adds visual appeal and a subtle sweetness. More than just a cocktail, this drink can evoke memories of enjoying frozen treats on a hot day, reinforcing how we can use memories as templates to sustain flavors even in the absence of smells.

Ingredients

60 mL (2 oz) vodka
30 mL (1 oz) fresh lemon juice
3 scoops lemon sherbet
120 mL (4 oz) prosecco
2 maraschino cherries (Luxardo brand)
2 bar spoons maraschino cherry syrup
2 rectangular strips of a preserved lemon rind (see note), remove any pulp

Instructions

In a cocktail shaker filled with ice, combine vodka and lemon juice. Shake for 10 s. Strain into a blender, then add lemon sherbet and prosecco. Blend on high for 10 s.

Divide equally between Nick and Nora.

Drizzle on the top of each cocktail the maraschino cherry syrup, then pierce the preserved lemon rind strip and maraschino cherry with a metal/wooden toothpick long enough to lay across the cocktail glass.

Preserved Lemons

Cut 6 lemons lengthwise, leaving ½ inch intact at the base so they stay whole. Rotate 90° and cut again to quarter, but not separate. Sprinkle lemons generously with salt in a large bowl, rubbing it in lightly. Pack into a sanitized 16 oz glass jar, pressing each lemon down to release juices. Drizzle with honey after each layer. Fill a ziplock bag with a 7% salt solution (7 g salt to 93 g water) and press it into the jar to remove oxygen. Seal and store in a dark, cool place for at least 2 weeks, or until the pith turns translucent. These are great for lots of applications and last forever in the fridge. Mix into mayo for a unique spread, or toss into stir fries or an omelet—these give you a salty, tangy element to any dish.

Modern Gold Rush

A riff off a riff of the classic whiskey sour, this cocktail leans into the richness of honey and spice, elevating the familiar with a higher sugar-to-acid ratio that enhances hidden autumnal flavors. While the drink remains balanced, bitters sharpen the sourness and temper the sweetness, while cinnamon from Goldschlager interacts with the alcohol burn, creating a dynamic trigeminal experience. Goldschlager also has gold flakes, making it much cooler than other cinnamon liquors (e.g., Fireball).

Ingredients

52.5 mL (1.75 oz) bourbon
22.5 mL (0.75 oz) rich honey syrup (see note)
15 mL (0.5 oz) fresh lemon juice
7.5 mL (0.25 oz) Goldschlager
2 dashes Angostura orange bitters

Instructions

In a cocktail shaker filled with ice, combine bourbon, honey syrup, lemon juice, Goldschlager, and bitters. Shake for 10 s to chill and dilute.

Strain into an old-fashioned glass over a large ice rock.

Express a lemon peel over the drink, and then place it as a garnish.

Rich Honey Syrup

Mix 250 mL (1 cup) honey with 83 mL (⅓ cup) water in a container. Stir until fully combined. Store in a tightly sealed container in the fridge for up to 2 weeks.

Dominick the Donkey (Nonalcoholic)

This nonalcoholic cocktail is a playful take on the classic Moscow Mule, reimagined with the depth and structure of a full-proof drink. Franklin Finney, a wonderful bartender who works at Public House in Knoxville, Tennessee, USA, created it. It works because it replicates the key physical elements that make cocktails satisfying: bite and body. The ginger beer delivers sharpness and carbonation, mimicking the familiar sting of alcohol on the palate. Meanwhile, the NA amaro provides body and balance with vegetable glycerin, which boosts viscosity for a rich, full mouthfeel, while its gentle sweetness offsets the bitter notes, much like a traditional amaro such as Averna.

Ingredients
22.5 mL (.75 oz) NA amaro (see note)
30 mL (1 oz) Chai Syrup (see note)
30 mL (1 oz) fresh lemon juice
105 mL (3.5 oz) ginger beer

Instructions
Build in a copper mug or Collins glass over ice.

Garnish with lime wedge.

NA Amaro
Toast 12 g guajillo (stemmed and mostly deseeded), 8 g ancho chilies (stemmed and mostly deseeded), 3 g (2 tsp) cloves, and 6 g (1 tbsp) allspice berries in large saucepan on medium heat until fragrant (1–2 min). Pay attention to the color change of the peppers and aromatics—keep away from any black, burned colors. Add 3 cups (700 mL) water, 1 lemon peel, 1 tangerine peel, 4 g (2 tsp) gentian, 1 g (1 tsp) wormwood, and 1 g (0.25 tsp) salt. Bring to a boil, turn off heat, then cover, and let steep for 1 h. Strain and discard the solids. Mix in 2 cups of vegetable glycerin to boost viscosity and give it a nice mouthfeel.

Chai Syrup
In a saucepan, combine 12 g (2 tbsp) of Harney & Sons chocolate Chai supreme loose-leaf tea and 125 g (0.5 cups sugar), then pour in 500 mL (2 cups) of boiling water. Stir until sugar dissolves, then cover and let it steep for 30 min. Strain through cheesecloth or a fine mesh sieve, then store in the fridge until use.

Hibiscus Mint Cooler (Nonalcoholic)

The cooler is a refreshing and vibrant nonalcoholic drink that balances floral and herbal notes with a touch of tartness. The striking deep red color from the hibiscus is visually stunning, while the mint adds a cooling layer. This drink is light, crisp, and perfect for a warm day. If you want to make an ice pitcher of this cocktail for a BBQ or picnic, make a fivefold concentration of everything, omit the shaking, and replace the seltzer with still water. If you're in the mood to imbibe, you can spike this cocktail by replacing 45 mL (1.5 oz) of the water with tequila.

Ingredients
30 mL (1 oz) mint-hibiscus syrup (see note)
15 mL (0.5 oz) fresh lime juice
135 mL (4.5 oz) cold soda water
Mint leaves

Instructions
In a cocktail shaker, combine mint-hibiscus syrup and lime juice. Add ice and shake for 10 s to chill and dilute. Strain into a highball glass filled with ice.

Tilt the glass at a 45-degree angle and slowly add soda water to preserve carbonation.

Garnish with fresh mint leaves.

Hibiscus Simple Syrup
In a saucepan, combine 250 g sugar and 250 g water, then add 15 g dried hibiscus flower and 25 g dried mint. Bring to a boil, turn off heat, then cover, and let it steep for 45 min. Strain through cheesecloth or a fine mesh sieve, then store in the fridge until use.

Tamarind Dream (Nonalcoholic)

This warm, spiced drink draws inspiration from the classic hot toddy, offering a soothing and layered experience without alcohol. The combination of lemon-ginger tea, tamarind, and citrus brings a balance of tangy, sweet, and spiced flavors, while the addition of clove-studded orange enhances both aroma and depth. Warm beverages like this are particularly pleasurable for those without smell, as the heat and spices engage trigeminal sensations, creating comforting warmth beyond flavor alone. This drink evokes nostalgia and seasonal coziness, reminiscent of fireside gatherings and winter traditions.

Ingredients

180 mL (6 oz) boiling water
1 pouch ginger tea
22.5 mL (0.75 oz) tamarind simple syrup (see note)
15 mL (0.5 oz) fresh lemon juice
1 orange slice
3 whole cloves
3 dashes orange bitters

Instructions

Press the three cloves into the flesh of the orange slice.

In an 8 oz mug, combine lemon juice, tamarind simple syrup, orange slice, and bitters. Stir to mix. Add the tea bag, then pour in boiling water.

Let steep for 4 min, then remove the tea bag.

Tamarind Simple Syrup

In a saucepan, combine 250 g sugar and 250 g water, then add 1 cup shelled tamarind or ½ cup tamarind paste. Bring to a boil, then use a potato masher to help separate the seeds from the tamarind flesh. Reduce heat, cover, and let sit for 2 min. Remove from heat and steep for 1 h. Strain through cheesecloth or a fine mesh sieve, removing seeds and pulp. Store in the fridge until ready to use.

Tamarind Dream (Nonalcoholic)

Guide Index

A. W. Fjaeldstad et al., *Rediscovering Flavor*,
https://doi.org/10.1007/978-3-032-08056-1

Index of Recipe

A. W. Fjaeldstad et al., *Rediscovering Flavor*,
https://doi.org/10.1007/978-3-032-08056-1